Patrick Moore's Practical Astronomy Series

T0222762

For other titles published in this series, go to
www.springer.com/series/3192

Measure Solar System Objects and Their Movements for Yourself!

John D. Clark

Springer

John D. Clark
323 Wootton Road
King's Lynn, Norfolk
PE30 3AX
United Kingdom
john.clark@finerandd.com

ISBN 978-0-387-89560-4   e-ISBN 978-0-387-89561-1
DOI: 10.1007/978-0-387-89561-1

Library of Congress Control Number: PCN Applied for

Printed on acid-free paper.

springer.com

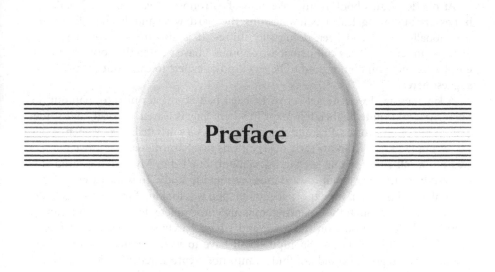

Preface

Until two-and-a-half years ago I had done no astronomy for 30 years. I had messed about with a home-made telescope as a kid, and taken an astrophysics class which barely mentioned the Solar System as part of my physics major, but that was it.

This course inspired me enough that I thought about studying astrophysics in graduate school, but I chickened out, figuring I would never get a job. So I did a PhD in semiconductor theory instead, only to find that there were no jobs in that either; so I washed up in medical device development, and pretty well forgot about astronomy.

In January 2006, my wife offered to buy me a 6-in. reflecting telescope. I knew so little about it that for almost a year I had the equatorial mount pointing south instead of toward Polaris. The "local" astronomy club met 70 miles away, so I rarely went, and in desperation I founded one locally. At last I had some friends to teach me the basics.

Gradually I noticed that although these folk knew much more about constellations, telescopes, and photography than me, I did look at the planets in a different way from them. I got little flashes of insight such as noticing that if Venus is at half phase, the angle between us, it, and the Sun must be a right angle, so the Sun, the Earth, and the Venus at that time made a right angle triangle, from which you can work out the distance to Venus. I do not think they were quite sure what to make of me and my mad ideas. One or two were unreceptive, but most club members seemed interested. I began to realize that I had a slightly different story to tell. This book is it.

You do not need a PhD in physics or astronomy to understand the Solar System. You can work out the basic layout of the Solar System from your backyard, even if you live in a brightly lit town like me.

At one level, this book is my collection of astronomy "war stories." I make no bones about having learnt a few lessons the hard way, and having discovered occasionally that I had been doing silly things. It is also the story of a journey from beginner to halfway competent astronomer. I have written the book like this to emphasize that you do not need to be any kind of expert to carry out the projects I suggest here.

A key message I want to get across is that I had no master plan to map out the Solar System. I quite deliberately leapt in where angels feared to tread, and sometimes took photos to see if I could detect any planetary or satellite movement at all. For example I knew from my early observations that Saturn does not move much. I was not at all sure I would see any movement, but I did. My activities gradually evolved from messing about toward more systematic study. To some extent I began to think that I had seen Jupiter often enough; and was looking for new things to do.

My journey through the Solar System is also incomplete. From England, Jupiter is practically on the horizon this year; and I have barely scratched the surface of what I could learn about it. So Jupiter will have to wait. Because the local light pollution is too great, I could not find Uranus nor Neptune reliably. Next autumn, I should be able to look for the larger asteroids.

There is another thing: my equipment is nothing special. I have a beat-up second hand 8-in. Newtonian reflector on a dual-axis driven EQ5 mount, as well as the 6-in. telescope my wife gave me. I have two Philips SPC900NC webcams, and a laptop PC. All the software I used is available for free, or an equivalent is, except K3CCD Tools^{TM}, which cost me US $50, and the software driver for my webcams, which of course came with them. Your equipment does not need to be fancy either.

To follow this book, you will undoubtedly need to know some high school geometry. Ideally you have done freshman year math in North America as part of a science or engineering course; or A-level maths or the Scottish or Irish equivalent in Europe. (I gave up on whether to write "math" or "maths" and will stick to mathematics.) If you picked up some basic calculus and geometry at night school on an apprenticeship of some kind, that should suffice. I have deliberately avoided using mathematical methods that require professional level skills: my methods are quite deliberately rather basic and downhome. This book is emphatically not for professional astronomers. Although I started out intending to write a book requiring basic calculus, I ended up using much less of it than I expected to. At a pinch you could skip those parts where I do use it.

I have, however, assumed that your mathematics are very rusty. Mathematical skills are not like swimming. You can do no swimming for years, and yet get straight in the water and swim again. A better analogy to mathematics is like letting your arm go numb if you lie in an awkward position. As the arm comes back to life, you will get pins and needles, which are no fun at all, but once they go away you will be fine. I think mathematics is a bit like that. At the start of the book, you may find the calculations uncomfortable, but once you find your stride they will get easier. Please allow a little time for this to happen; and do not worry. Almost every other reader will experience the same. Do not expect to read the book all in one go like an airport paperback. Take your time, think about it, take lots of breaks, and be prepared to have a pencil and paper handy to work through the calculations. Real calculations do not come in neatly packaged, same-size chunks like school textbook problems. They may only use textbook concepts, but some can be lengthier than simple textbook examples.

I have added Appendices A and B to help you get started. These are not primers: they are necessarily minimalist, and are there because most of you will not live in a house with a dozen mathematics texts. I am supposed to be writing a book about astronomy here, not mathematics. What I have tried to do is to provide all the algebraic manipulation. Most people (me included) hate seeing calculation steps missed out, or worse yet left as an exercise for the reader. It is like having to navigate in the pre-GPS era with one missing road sign at a busy intersection. You are stymied.

I am a great believer in the evidence-based approach. That is a key part of the ethos of this book. You can look up the distances of the planets from the Sun on the Internet in 10 min. But how do you know that these distances are close to right? What I want to show you is how you can check the approximate truth of the planetary distances that every astronomer "knows." I find it much more satisfying to know where a fact comes from. In the first instance, the thing is to get a rough idea. That is what I aim to give you. It is good science to get a rough idea before you look for a detailed one. If you want to send a rocket to Mars though, you will need more sophisticated methods than I offer.

Where did I get my methods from? I looked in many books, notably those by French; Danby; Murray and Dermot; Ferguson and Tatum[1] listed in the bibliography. Some I knew from my college days, notably my freshman mechanics course. I concluded that the methods given in the celestial mechanics texts would be beyond most amateurs, because they focus on elliptical rather than circular orbits. The only one who treats circular orbits is Tatum, who points out that for most planets the orbits are almost circular. Even his treatment is not suitable for our purposes. First, he leaves most of the analysis as an exercise for the reader (see above). Second, his method for obtaining planetary distances requires knowing things you have no hope of measuring. I do not think he has tried his method in anger, or he too would have found this out. I only noticed because I did try his method. So in the end, I made the methods up. I do not believe for one second that I am the first person to analyze circular orbits. I am sure many others have, but I could not find where they wrote up their analyses. I was proud of my cleverness in one or two places, but mostly I think I was just doing my job as a physicist.

I have seen "Monte Carlo" statistical methods used for many things, e.g., in Wall and Jenkins' *Practical Statistics for Astronomers* or the *find_orb* orbit determining software (http://www.projectpluto.com). I did not copy my method for performing least squares analysis from anywhere. It is a technique a professional physicist like me ought to be able to apply, just like a dentist should know how to drill teeth. I think Monte Carlo methods are accepted to be a last resource of the mathematically

---

[1] Tatum, J.B. (2007) Celestial Mechanics, online book found at http://www.astro.unic.ca/~tatum

desperate, a state in which I most certainly was. In the sense that they require a lot of number crunching, they are not very efficient; but for our purposes their inefficiency is of little consequence. Your PC should handle these methods easily. They may take up to an hour each per planet on a 10-year-old computer, and much less on a newer one, but what is the big deal about that? You can spend a lot more time than that sitting by your telescope waiting for clouds to clear. The key advantage of Monte Carlo methods is that they are simple and easy to understand.

I firmly believe that if you do not understand a statistical method, you should not be using it. For example, in business school, we were taught a technique for hypothesis testing (the chi-square method) without being told how it works. That is outrageous: to proclaim the truth or otherwise of a hypothesis without knowing why is no better than to assert that retrograde planetary motion causes bad karma. I have provided an appendix deriving the statistical methods I use from first principles. You will definitely need calculus to follow it. If that is beyond you, you will survive by skipping this appendix. However, you will have to accept that random scatter in measurements tends to be distributed in a "bell curve" with most measurements near the average. You will also have to take my word for it that when fitting lines to data, this implies that you should use the method of least squares. Your high school geometry and algebra will enable you to implement this method even if you do not understand where it comes from.

Since I work in new product development, I am now also a qualified engineer, and have freely borrowed ideas from that world. In particular, I have never seen or heard of anyone using Computer-Aided Drafting (CAD) software to make measurements from astronomical photographs. You can download CAD software for free. This software usually allows you to import digital images, and has dimensioning tools to allow lengths and angles to be very quickly and accurately added to engineering drawings. What I did was to use the software to draft circles, and occasionally ellipses, freehand around the celestial objects, and use the dimensioning tools to make the measurements. There is professional astronomical software around with not dissimilar capabilities, but it only runs on computers with Unix or Linux operating systems. There is a package called Astro Art (http://www.msb-astroart.com) which does it all for you, but as far as I can make out it is a "black box" that does not tell you what it did. By now you should be able to guess whether I approve of that.

If you enjoy this book I will have been successful, though I would not of course know it. The thing is to have a go. Although my recipes have all been tested, and I report how good they are, you need not follow them precisely if you do not want to. I primarily want to feed you a few ideas.

If someone is inspired to contact me at john.clark@finerandd.com with ideas about observing the Solar System, or even write a book, showing how they learned lots of fun stuff about Solar System objects, I would be delighted.

King's Lynn, UK   John D. Clark

Acknowledgments

I am long enough in the tooth to owe intellectual debts to many people. It starts with my parents, of course. The inspiring teaching of my school physics teacher, the late Brian Salthouse, helped shape my decision to study science – I could just as easily have become a linguist or a classicist. A few of my undergraduate teachers were terrible, and should have been fired, but those who were good were very, very good. In an astronomical context I would single out Gary Gledhill, who taught my freshman mechanics course. To this day I regard it as one of the great intellectual experiences of my life. I have used quite a bit of the material from that course here. I would also like to thank Peter Williams, who taught me Astrophysics and was my personal tutor throughout my bachelor's degree; and one person who I think would be surprised: Wladimir von Schlippe. He set me up for life with an undergraduate mathematics course. At the time I was critical with the certainty of youth, because he was not flashy and sparkling. But so what? His teaching was clear and thorough, and I learnt as much mathematics from him as from anyone. I still feel ashamed that I was so critical of him.

Over the years, I owe debts to other people from whom I learnt a lot about science: the late Paul Butcher, my PhD supervisor; Phil Taylor, my first boss; Bill Potter, my next one, who hired me a second time 12 years later; another boss Mark Wickham, who also hired me twice in two different firms; and three friends who over the years have challenged me with tough questions: Charles Jenkins, Nick White, and Jim Franklin.

I must also thank the people who read and criticized the manuscript, primarily my wife and daughter, but also Paul Millar and Gary Wassell. Paying my daughter £5 ($10) for every mistake she found proved to be expensive. My Dad provided the hand drawings. Nevertheless only one person is responsible for any remaining errors: me.

Of course the debt I owe my wife, daughter, and parents for love and support is beyond my capacity for repayment.

I have learnt a lot of observational tricks from various astronomy club members, including Sue Napper from Norwich Astronomy Society; and Freddy Rice, Adrian King, Trevor Nurse, and Darren Sprunt from West Norfolk Astronomy Society.

Now I have had to learn a new art: how to put a book together. The staff at Springer, John Watson, Maury Solomon, and Turpana Molina, have guided me patiently through this process.

Contents

About the Author

John Clark's wife bought him a telescope in 2006. He was immediately and badly bitten by the astronomy bug and started taking photographs a year later. Feeling rather isolated, he founded the West Norfolk Astronomy Society, and was recently awarded a grant by the Institute of Physics for astronomy outreach activities. An engineering physicist by profession, he brings insights from both engineering and physics into his astronomy in a unique combination. He lives in King's Lynn, England, with his wife and daughter; and with his parents nearby.

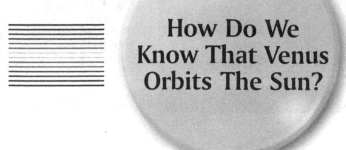

CHAPTER ONE

How Do We Know That Venus Orbits The Sun?

Venus exhibits phases like the Moon. It is a lot further away, so it looks a lot smaller than the Moon. The phases are not visible to the naked eye, except to a very few sharp-eyed people; and were unknown until the advent of the telescope about 400 years ago.

You cannot see the phases very well in 10 × 50 binoculars, but my 6 in. *f*/5 Newtonian telescope shows them very nicely. At least it did in late winter 2006. The crescent phase was unmistakable. But when it first appeared in the evening sky around January 2007, I was very puzzled because I had seen the phases so easily the previous winter just before sunrise. I could not see any phases. The darn thing was just a bright blob even at maximum magnification.

I told my wife, not very confidently, that it must be much further away than a year ago. Was I right?

It was about this time that I first tried my hand at photography with a webcam. My first photos were not a pretty sight. It took me a couple of months to get around to photographing Venus (Fig. 1.1). By this time I had bought an 8 in. *f*/6 telescope through e-Bay, which was a big improvement on the 6 in., mainly because it has an electric focuser and an EQ5 mount, which is much stiffer than the EQ2 on the 6 in. model. The better optics and the dual axis drive certainly do not hurt, but to me they are secondary benefits for photography. Anyway, it was April 10th by the time I first had a go at photographing Venus. I could now resolve its shape. It was gibbous.

I cannot claim to have been very systematic about my photography. The weather was not helpful. Nor, at this point, was my rather variable photographic technique. Nevertheless I watched it grow bigger, and watched the phase wane. By late May it had reached half-phase.

J.D. Clark, *Measure Solar System Objects and Their Movements for Yourself!*,
Patrick Moore's Practical Astronomy Series,
DOI: 10.1007/978-0-387-89561-1_1, © Springer Science + Business Media, LLC 2009

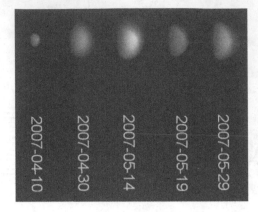

Fig. 1.1. The author's own photographs of the phases of Venus, taken with the same telescope at the same magnification, over a 7-week period.

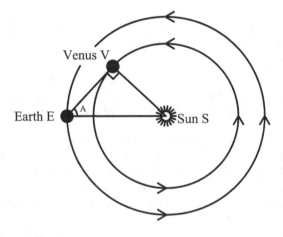

Fig. 1.2. If Venus is at half-phase, the Earth-to-Venus line must be at right angles to the Sun-to-Venus line. Together with the Sun-to-Earth line, these lines form a right-angle triangle. We shall call the vertices of this triangle S, E, and V. When the angle SVE is a right angle and Venus' phase is waning, the configuration of the planets is known as the Greatest Eastern Elongation of Venus.

At this point I had a brainwave. If the planet was at half-phase, the angle between the Earth-to-Venus line and the Venus-to-Sun line must be a right angle.

"So what?" you ask.

Well, from Fig. 1.2 we see that the Earth, the Sun, and the Venus make the right-angle triangle SVE, where S, E, and V mean "Sun," "Earth," and "Venus," respectively. The right angle is at V, as shown in Fig. 1.2. The properties of right angle triangles are well documented. I can use them to estimate the distances EV and VS if I know the distance ES. To a first approximation, the distance ES is known as one

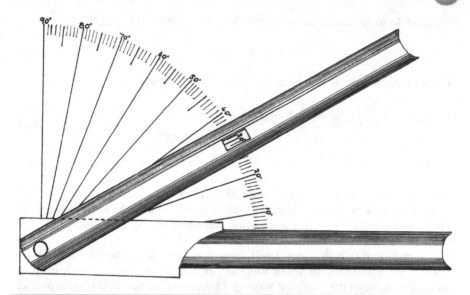

Fig. 1.3. My home-made device for measuring the angle between heavenly bodies. It consists of a cardboard backing onto which is fastened an angle scale drawn with computer-aided drafting (CAD) software. Thumb tacks are used to attach two cardboard coat hanger inserts of the type dry cleaners use to hold long pants in place.

astronomical unit (AU). More accurately, 1 AU is an average Earth-Sun distance: the orbit is of course not a perfect circle.

Not quite: I need one more piece of information – I need to know the angle A in Fig. 1.2. This means I need to measure it. I tried using rulers and a protractor, but got nowhere. It was time for a bit of second childhood while I made a measuring device.

Figure 1.3 hows the offending item. It consists of a cardboard backing onto which is fastened an angle scale drawn with computer-aided drafting (CAD) software. Thumb tacks are used to attach two cardboard coat hanger inserts of the type dry cleaners use.

The trick was to wait until just before sunset, when the Sun was not blinding, and place my eye where the two cardboard inserts met. I pointed one cardboard insert at the Sun, the other at Venus. The v-shaped nature of the inserts made the pointing a little easier. I managed this twice on the evening of 29 May, and recorded angles of 46° and 47°. Although my device worked, it was not exactly user-friendly – measurement was difficult.

Incidentally, the planetary arrangement shown in Fig. 1.2 is known as the "Greatest Eastern Elongation" of Venus. This is because if the angle A were to get any bigger than this, the line EV would miss the orbit of Venus completely. Therefore angle A cannot get any bigger than at Greatest Eastern Elongation.

We now do some simple trigonometry. Taking the angle A in Fig. 1.2,

$$\sin A = \frac{SV}{SE}. \qquad (1.1)$$

But we have measured A to be about 46.5°. So (1.1) becomes

$$\sin A = \sin 46.5 = \frac{SV}{SE} = 0.725; \qquad (1.2)$$
$$\therefore SV = 0.725\,SE \approx 0.73\,AU.$$

In (1.2), the symbol "$\approx$" means "is approximately equal to." I now have a distance estimate: the distance from the Sun to Venus is between 0.72 and 0.73 AU. Bakich[2] reports values between 0.7184 and 0.7282 AU, with an average of 0.7233. Given the crudity of my measurement, I did better than I deserve. One of the weaknesses of the scientific measurement is that it gets awfully tempting to stop experimenting when you find the answer you were looking for.

Far from falling into this trap, I will show you later how I double-checked my measurement. I waited until Venus was next at half-phase, but waxing instead of waning. This is called the "Greatest Western Elongation" of Venus. If I could predict the time at which this happened, I would have understood the relative orbits of Earth and Venus correctly. In particular, a prediction of the time at which the planet is next at half-phase can be checked with my telescope.

To make this prediction, I need to show you a little bit about how planetary orbits work. Since neither the Earth nor Venus has a very elliptical orbit,[2] I can get away with assuming that these two planets have circular orbits. This assumption simplifies the calculations from graduate-school level to freshman year science or engineering (A-level mathematics or physics in Britain).

Circular Motion

The physics of circular motion was first solved by Christiaan Huyghens in 1688.[3]

First, I will show you how to work out this physics in a way that uses no calculus.

Imagine a point which goes around a circle of radius r with constant speed v. Such a point is shown in Fig. 1.4. Just as the radius r keeps its magnitude but goes around in a circle, so does the direction of v.

I am going to define a unit of angular measure called a "radian." Figure 1.5 shows how the angle is one radian when the length of the arc between two radii is equal to the length r of the radius. Since the length of the entire circumference is $2\pi r$, one complete circle contains 2π radians. In other words, 2π radians = 360°, or 1 radian $\approx 57.296°$.

If in Fig. 1.4 the time taken to complete one revolution is T, then the time to complete one radian is $T/2\pi$. We therefore say that the angular velocity of the rotating point is

$$\omega = \frac{2\pi}{T}, \qquad (1.3)$$

where the Greek letter ω (lower case omega) is the traditional symbol for angular velocity in radians per second.

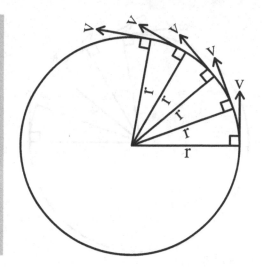

Fig. 1.4. A point on the circumference of a circle of radius *r* rotates around it with uniform speed *v*. Since the velocity is always perpendicular to the moving radius, it too goes around in a circle.

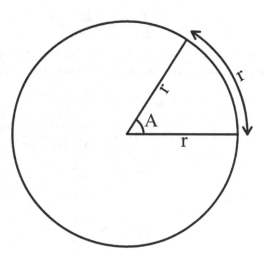

Fig. 1.5. If the length of the arc between two radii is equal to the length *r* of the radius, the angle *A* is said to be equal to one radian. Since the length of the entire circumference is $2\pi r$, one complete circle contains 2π radians. In other words, 2π radians = 360°.

What is the value of T? Speed is the distance traveled in unit time.

$$\text{Speed} = \frac{\text{Distance}}{\text{Time}};$$

$$\therefore \text{Time} = \frac{\text{Distance}}{\text{Speed}};$$

$$\text{i.e.,} \quad T = \frac{2\pi r}{v}.$$

(1.4)

We can also write T in terms of ω. Substituting (1.3) into (1.4) gives

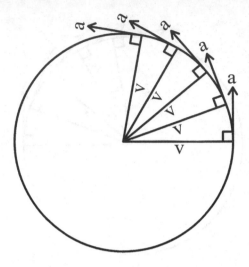

Fig. 1.6. The speed of the rotating point does not change, but its velocity does. This is because velocity has direction as well as magnitude. It is possible to imagine the velocity as the "radius" of a "circle of velocity."

$$\frac{T}{2\pi} = \frac{1}{\omega} = \frac{2\pi r}{2\pi v} = \frac{r}{v}. \tag{1.5}$$

There is a striking analogy between the circles of Figs. 1.4 and 1.6. In both cases, the time to complete once circumference is T. By direct analogy with (1.4),

$$\text{Acceleration} = \frac{\text{Speed change}}{\text{Time}};$$

$$\text{Time} = \frac{\text{Speed change}}{\text{Acceleration}}; \tag{1.6}$$

$$\text{i.e.,} \quad T = \frac{2\pi v}{a}.$$

By analogy with (1.5)

$$\frac{T}{2\pi} = \frac{1}{\omega} = \frac{2\pi v}{2\pi a} = \frac{v}{a}. \tag{1.7}$$

Why did I do this? Because I can use (1.7) and (1.5) to figure out what the acceleration of a point going around in a circle is. Watch this. First I use (1.7) to get one formula for the acceleration, then I use (1.5) to get rid of the velocity term, which I do not really want.

$$a = \omega v = \omega(\omega r) = \omega^2 r. \tag{1.8}$$

A very important law in dynamics is Isaac Newton's second law of motion. This law is given in his book "Mathematical Principles of Natural Philosophy," [4] although even the English translations of this Latin work look very old fashioned and strange to our eyes. In more modern language, Newton's second Law states that

$$F = ma, \tag{1.9}$$

or Force = Mass × Acceleration. Mass is measured in such units as tons, pounds, and kilograms.

When I worked out (1.8), I just talked about a geometric point. I never said that there was anything there. Now, I am going to imagine that there is something there. It will be just a small something, with mass m. In particular, let us imagine that its size is small compared to the radius r. Then I combine (1.8) and (1.9) to give

$$F = ma = m\omega^2 r. \tag{1.10}$$

This is the force that must be applied to the mass m to get it to go around in a circle. It is known as the "centripetal force" because "centrum" means center and "petere" means "to move toward" in Latin. I am sure the Romans talked about petering toward the centrum all the time. What else was there to do back then? However, we got the last laugh: it is Latin that has "petered" out.

Why must a centripetal force be applied? Because Newton's law, (1.9), would give zero acceleration without a force, and zero acceleration means going at the same speed in a straight line forever. This "natural" state of unforced motion was so surprising in Newton's time that he made it a separate law of motion, which we now call Newton's first law of motion: "a body continues at rest or in uniform motion in a straight line unless it is acted upon by an external force."

When I first learned to calculate the acceleration this way, at the age of about 15, I have to confess that the argument did not impress me. So I am going to show you how to derive (1.8) with calculus, just to show you how much easier calculus makes your life.

Consider the point P in Fig. 1.7. It has coordinates $(x, y) = (r\cos[A], r\sin[A])$. However, if the point P is going round with angular velocity ω, then it follows that

$$A = \omega t, \tag{1.11}$$

So that

$$(x, y) = (r\cos(\omega t), r\sin(\omega t)). \tag{1.12}$$

If you differentiate position with respect to time, you get velocity, which you can think of as "rate of change of position with respect to time." Also, if you differentiate velocity with respect to time, you get acceleration, which you can think of as "rate of change of velocity with respect to time." Therefore if you differentiate position with respect to time twice, that is tantamount to differentiating first position then velocity with respect to time, so you end up with acceleration.

In case you have forgotten, the formulae for differentiating sines and cosines are

$$\frac{d\sin(\omega t)}{dt} = \omega r \cos(\omega t)$$

$$\text{and} \quad x\, \frac{d\cos(\omega t)}{dt} = -\omega r \sin(\omega t), \tag{1.13}$$

provided that ω is in radians per second. If you use degrees, the formulae become more complicated. This is a major benefit of using radians.

The velocity coordinates are therefore

$$\left(\frac{dx}{dt}, \frac{dy}{dt}\right) = (-\omega r \sin(\omega t), \omega r \cos(\omega t)), \tag{1.14}$$

and the acceleration coordinates are

$$\begin{aligned}\left(\frac{d^2x}{dt^2}, \frac{d^2y}{dt^2}\right) &= (-\omega^2 r \cos(\omega t), -\omega^2 r \sin(\omega t))\\ &= -\omega^2 (r \cos(\omega t), r \sin(\omega t))\\ &= -\omega^2 (x, y).\end{aligned} \tag{1.15}$$

In the last step, (1.12) has been used. There are no appeals to you to imagine a circle whose radius is not a distance but a velocity. You just have to know how to differentiate.

Also, in Equation (1.15) a minus sign has appeared. This is not wrong: it indicates that the acceleration is from the circumference toward the center, whereas the radius goes from the center toward the edge. This is another benefit to using calculus – it gives a way to keep track of direction.

Actually, I have snuck another mathematical concept in. By using coordinates, I wrote the radius in such a way that its direction is accounted for as well as its magnitude. To cut a long story short, radius, velocity, and acceleration each have direction as well as magnitude, and satisfy certain of other conditions, which need not concern us, to make them "vector" quantities. Writing things in coordinates is just about the easiest way to handle vectors.

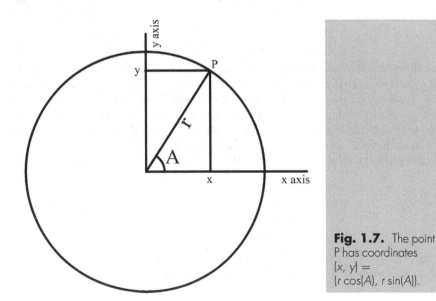

Fig. 1.7. The point P has coordinates $(x, y) = (r \cos(A), r \sin(A))$.

Newton's Law of Gravity

In his "Mathematical Principles"[4] Newton also showed how one law of gravity describes the behavior of both celestial and earthly bodies. Imagine two bodies of masses M and m, separated by a distance r. They attract one another with a force F given by

$$F = \frac{GMm}{r^2}. \tag{1.16}$$

When orbiting, the two bodies of masses M and m actually orbit about their center of mass. If M is vastly greater than m, this center of mass is approximately at the center of the mass M. The Sun is a third of a million times more massive than either Earth or Venus, so I am going to assume that both planets orbit about the center of the Sun.

By the way, M has to be an awful lot greater than m for this approximation to work. For example, even though the mass of Jupiter is just under a thousandth that of the Sun, the center of mass of the Sun and the Jupiter is outside the photosphere of the Sun. If you are interested, the relevant formulae can be found in Murray and Dermott's book "Solar System Dynamics".[5]

If you want to write (1.16) in vector form, you can. You have to insert a vector of unit magnitude (see Appendix A) that points along the line joining M and m, and insert a minus sign to indicate that the force is attractive, i.e., it points toward the coordinate system origin at the center of mass.

Application to Circular Orbits

The centripetal force required to keep the planet in an orbit of radius r is provided by gravity. It is as simple as that. The orbit is stable if the forces in (1.8) and (1.16) are one and the same:

$$F = \frac{GMm}{r^2} = m\omega^2 r. \tag{1.17}$$

This equation can be rearranged. First, the mass m cancels out. Second, we combine the radii:

$$\frac{GM}{r^3} = \omega^2 = \left(\frac{2\pi}{T}\right)^2. \tag{1.18}$$

Therefore

$$\frac{GM}{4\pi^2} = \frac{r^3}{T^2} = \text{constant}. \tag{1.19}$$

Equation (1.19) is a way to test whether a given heavenly body is in fact in an orbit. This law, which was discovered by Johannes Kepler,[6] works for elliptical as well as circular orbits, provided r is replaced with a suitable quantity. It is known in the scientific literature as Kepler's third law of planetary motion.

Back to Venus and the Earth

In (1.2), I gave my measured value of r for Venus:

$$r_{\text{Venus}} = 0.73\,\text{AU}. \tag{1.20}$$

By definition

$$r_{\text{Earth}} = 1\,\text{AU}. \tag{1.21}$$

Now I am going to substitute these values into (1.19):

$$\frac{r_{\text{Earth}}^3}{T_{\text{Earth}}^2} = \frac{r_{\text{Venus}}^3}{T_{\text{Venus}}^2} = \text{constant}. \tag{1.22}$$

Hence

$$\frac{r_{\text{Earth}}^3}{r_{\text{Venus}}^3} = \frac{T_{\text{Earth}}^2}{T_{\text{Venus}}^2} = \frac{1^3}{0.73^3} = \frac{1}{0.389017}. \tag{1.23}$$

Rearranging

$$\frac{0.389017}{T_{\text{Venus}}^2} = \frac{1}{T_{\text{Earth}}^2}$$
$$\therefore 0.389017\, T_{\text{Earth}}^2 = T_{\text{Venus}}^2 = 0.389017 \times 1^2 = 0.389017; \tag{1.24}$$
$$\therefore T_{\text{Venus}} = \sqrt{0.389017} = 0.623\,\text{years} = 227.8\,\text{days}.$$

The published value of the time for Venus to orbit is 0.615 years or 224.7 days.[2] My measured value is accurate to within 1.3%. The approximation that the orbits of Earth and Venus are circular is not looking too bad. Let us see if I can actually *measure* the orbital period.

Before I try to predict the time of the Greatest Western Elongation of Venus, I have a confession to make. As Fig. 1.8 shows, I jumped the gun when I assumed that the

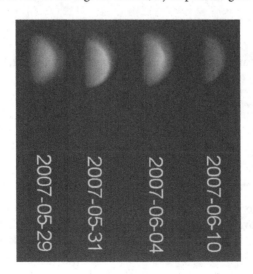

2007-05-29 2007-05-31 2007-06-04 2007-06-10

Fig. 1.8. The author's own photographs of the phases of Venus taken around the time of Greatest Eastern Elongation.

Greatest Eastern Elongation of Venus had already occurred by May 29, 2007. Subsequent photos showed that it probably occurred between May 31 and June 10. I cannot tell more precisely than that. The main reason I did not have a daily photo was poor weather, although my evening diary was not without nonastronomical engagements that week, not least my daughter's birthday on June 8. I will therefore assume that the actual date was June 5 2007 ± 5 days.

How to Predict the Greatest Western Elongation Date

First, let me convert (1.23) from orbital periods into angular velocities using (1.5)

$$\frac{\omega_{\text{Venus}}}{\omega_{\text{Earth}}} = \frac{2\pi/T_{\text{Venus}}}{2\pi/T_{\text{Earth}}} = \frac{T_{\text{Earth}}}{T_{\text{Venus}}} = \frac{1}{0.624} = 1.603. \tag{1.25}$$

But we know that $\omega_{\text{Earth}} = (2\pi/365.25)$ radians per day.

Let us define a coordinate system with its origin at the center of the Sun. Angles are measured from the x-axis, with the positive direction being anticlockwise toward the y-axis. There is no reason for defining angles this way: it is merely the custom. We need to think a little bit about how this coordinate system is moving. After all, the planets orbit, and the Sun rotates about its own axis roughly once a month, with some parts rotating faster than others. What's more, the Sun is traveling in an orbit about the center of our Galaxy. Clearly our coordinate system needs to move around the Galaxy with the Sun. But we want it not to rotate. This begs the rather awkward question: relative to what might this coordinate system rotate? For the few months that I want to follow the motion of planets, it will be good enough to postulate that this coordinate system does not rotate relative to the stars, ignoring the few closest neighbors that have large proper motions. If we try to define a better standard of nonrotation, we will open a Pandora's Box.

If a planet, be it Earth, Venus, or any other, is orbiting the Sun with angular velocity ω, then the angle from the x-axis to it is $\omega(t-t_0)$, where t_0 is a time at which the planet crossed the x-axis. t_0 is different for each planet. One of the curious features of the Newtonian system of mechanics is that there is no "time zero." We must arbitrarily choose one. You can do this with a stop watch. It is a very mundane thing to do. You start the stop watch at the time you want to be "time zero." The smart thing to do is to choose the zero time that makes your life as simple as possible. Sometimes you only know this time after you have tried the calculation, and end up redoing the mathematics. I could not possibly comment on whether this happened here.

Anyway, a good time zero is the last time before late May 2007 that the Earth and Venus were at the same angle with respect to the x-axis. This situation is shown in Fig. 1.9.

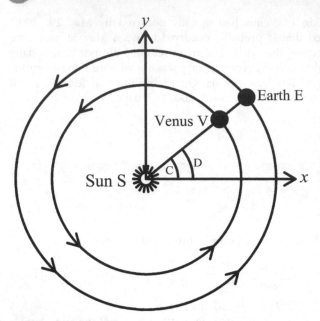

From Fig. 1.11 we see that at Greatest Eastern Elongation

$$\text{Angle } C = \text{Angle } D - (90 - 46.5)° \quad \text{or}$$
$$\omega_{\text{Venus}} t_{\text{GEE}} = \omega_{\text{Earth}} t_{\text{GEE}} - (90 - 46.5)°, \tag{1.26}$$

where t is time and the subscript GEE means Greatest Eastern Elongation. We are using the idea that the angle through which a planet has orbited since time zero is given by

$$\text{Angle} = \omega t. \tag{1.27}$$

At Greatest Western Elongation

$$\omega_{\text{Venus}} t_{\text{GWE}} = \omega_{\text{Earth}} t_{\text{GWE}} + (90 - 46.5)°, \tag{1.28}$$

where the subscript GWE means Greatest Western Elongation. This situation is shown in Fig. 1.10.

Substituting for ω_{Venus} from (1.25) gives

$$1.603 \omega_{\text{Earth}} t_{\text{GEE}} = \omega_{\text{Earth}} t_{\text{GEE}} - (90 - 46.5)° \tag{1.29}$$

and

$$1.603 \omega_{\text{Earth}} t_{\text{GWE}} = \omega_{\text{Earth}} t_{\text{GWE}} + (90 - 46.5)°. \tag{1.30}$$

Equations (1.29) and (1.30) can be rearranged and subtracted from one another as follows. First let me rearrange them:

$$(1.603 - 1) \omega_{\text{Earth}} t_{\text{GEE}} = 0.603 \omega_{\text{Earth}} t_{\text{GEE}} = -(90 - 46.5)° \tag{1.31}$$

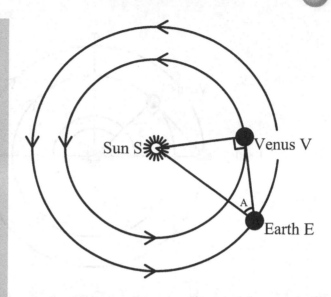

Fig. 1.10. By the time Venus has reached its Greatest Western Elongation, both it and the Earth have moved along their orbits compared to Fig. 1.11. Venus moves the faster, being nearer to the Sun, and therefore experiencing stronger solar gravity. This diagram shows how the angle SVE is once again a right angle, and Venus is at half-phase, albeit waxing this time.

and

$$(1.603 - 1)\omega_{\text{Earth}}\, t_{\text{GWE}} = 0.603\omega_{\text{Earth}}\, t_{\text{GWE}} = +(90 - 46.5)^\circ. \qquad (1.32)$$

Now let me subtract (1.31) from (1.32):

$$0.603\omega_{\text{Earth}}(t_{\text{GWE}} - t_{\text{GEE}}) = (90 - 46.5)^\circ + (90 - 46.5)^\circ = (180 - 93)^\circ$$
$$= 87^\circ = \frac{87^\circ \times \pi}{180}\ \text{radians} \qquad (1.33)$$

I can substitute the value $\omega_{\text{Earth}} = (2\pi/365.25)$ radians per day into (1.33) to give

$$0.603\omega_{\text{Earth}}(t_{\text{GWE}} - t_{\text{GEE}}) = 0.603\frac{2\pi}{365.25}(t_{\text{GWE}} - t_{\text{GEE}})$$
$$= \frac{87^\circ \times \pi}{180}\ \text{radians}. \qquad (1.34)$$

Therefore

$$\frac{87^\circ \times \pi}{180}\ \text{radians} = 0.603\frac{2\pi}{365.25}(t_{\text{GWE}} - t_{\text{GEE}})$$
$$\therefore (t_{\text{GWE}} - t_{\text{GEE}}) = \frac{365.25 \times 87 \times \pi}{0.603 \times 2 \times \pi \times 180} = 146.3\ \text{days}. \qquad (1.35)$$

If GEE was on June 5 2007 $\pm$ 5 days, (1.35) implies that

$$t_{\text{GWE}} = \text{October 29 2007} \pm 5\ \text{days}. \qquad (1.36)$$

At this time, Venus would again be at half-phase, but waxing not waning.

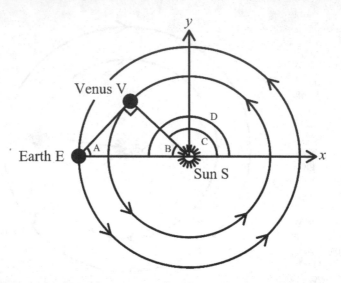

Fig. 1.11. The situation at Greatest Eastern Elongation (GEE). All angles are measured anticlockwise from the *x*-axis. Let the time of GEE be t_{GEE}. Then the angle *C* can be written as $\pi - B$.

Checking the Date of Greatest Western Elongation

So what actually happened? Figure 1.12 shows the sequence of photos I took. All the waning ones up to July 6 were taken in the evening, and all the waxing ones from September 9 were taken in the morning. I did not get any in between because Venus was very low in the sky. My backyard is surrounded by trees. The view to the north-west is particularly blocked, so the waning crescent was particularly hard to capture on a summer evening. The main block to the south-east is the houses opposite. But worse, much worse than that was the weather. The inferior conjunction, when Venus was between the Earth and the Sun, would have taken place halfway between GEE and GWE, on about August 17. From then until September 9 there were very few clear mornings. I remember one morning when I took my second-ever photo of Mars through thin cloud. That was about all I got during those 23 days.

Figure 1.13 shows that the actual date of Greatest Western Elongation was after October 22 but before November 6. In fact careful examination of the November 3 photo (Fig. 1.14) shows that Venus was already barely gibbous. The nearest photo to the actual Greatest Western Elongation was that of October 31. This is completely consistent with the prediction of (1.36).

Incidentally, whenever I tell anyone what time I took the morning photos, even astronomers give me horrified looks. Dragging yourself out of bed for pre-dawn photography has obvious drawbacks, but it does have one advantage. Astronomy was off the menu in June for my daughter's birthday. But at 5 A.M. on October 22, my wife's birthday, she did not care what I was doing. And later that day, I got to enjoy the party without thinking about missing key photo opportunities for my project.

2008-01-30
2008-01-05
2007-12-09
2007-12-01
2007-11-24
2007-11-17
2007-11-06
2007-11-03
2007-10-30
2007-10-22
2007-10-21
2007-10-18
2007-10-14
2007-10-05
2007-09-30
2007-09-23
2007-09-16
2007-07-06
2007-06-27
2007-06-20
2007-06-10
2007-06-04
2007-05-31
2007-05-29
2007-05-19
2007-05-14
2007-04-30
2007-04-10

Fig. 1.12. The author's own photographs of the phases of Venus, taken with the same telescope at the same magnification, over a 7-month period. My photographic technique improved as I shot the pictures. There is no shame in this. We all have to learn somehow. Your technique will no doubt also improve with experience.

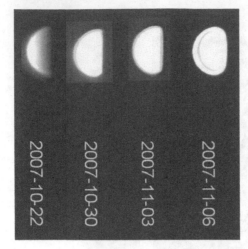

2007-10-22   2007-10-30   2007-11-03   2007-11-06

Fig. 1.13. Detail from Fig. 1.12 showing that Greatest Western Elongation happened after October 22 and before November 6.

Fig. 1.14. Detail from Fig. 1.13 showing that Greatest Western Elongation had already happened by November 3.

On 30 October, the weather was good enough for me to repeat my angle-measurement experiment. This time I managed three measurements, of 47°, 48°, and 48° again. If I repeat the calculation from (1.26) through (1.36), using the average of my five values 46°, 47°, 47°, 48°, and 48°, viz., 47.2°, I estimate the date of Greatest Western Elongation to be October 26 2007 ± 5 days. This is still consistent with it being known to be after October 22 but before November 3.

Alternative Hypothesis: Venus Orbits Earth

In ancient times, various theories were produced in which the planets, the Moon and the Sun all orbited the Earth. The most noted proponent of such a theory was Ptolemy,[7] who lived around 100 C.E. I will now show how such a theory is rendered implausible by telescopic observations, which of course were not available to Ptolemy.

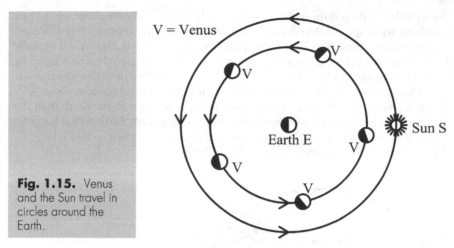

Fig. 1.15. Venus and the Sun travel in circles around the Earth.

If, as in Fig. 1.15, Venus and the Sun were to travel in circles around the Earth, we would observe the following. The phases of Venus could be either crescent or gibbous, as Fig. 1.15 shows. We observe this. Venus would always be about the same size. We do not observe this – my own photographs shown in Fig. 1.12 disprove it. There would also be no angles of Greatest Eastern or Western Elongation: Venus might be visible at any time of the night. In practice it is only visible within about 2 h of sunset or sunrise. Hence the scenario of Fig. 1.15 is not consistent with what we observe.

The technique I am using here is called "proof by counter-example." The idea is that if a given scenario is shown to "predict" at least one disprovable observation, that one counterexample is enough to disprove the whole scenario. It is a very elegant technique.

There is only one problem: it only works perfectly in mathematics, where things can be proved or disproved beyond reasonable doubt; and proofs can be repeated and checked ad nauseam. In experimental science, it is always possible to dispute the validity of the counterexample. It could be claimed, for example, that the leftmost photo of Venus in Fig. 1.12 is much smaller; so that in fact I had photographed Mercury by mistake. I cannot turn the clock back and check this, though I can look in tables and find that Mercury was a morning star on April 10 2007. In real research problems, there may not be a set of tables I can look in, because by definition real research has never been done before. This difficulty of proof by counterexample is acute in such fields as Cosmology and Evolutionary Biology, where we have no way to turn the clock back and see if our theories are right or wrong. Sadly there is much misunderstanding of this point.

Sometimes, an idea takes an awful lot of killing by counterexample. Percival Lowell's "discovery" of canals on Mars was a classic example. This idea was not taken seriously by scientists for long, but it was not expelled from public consciousness until the first space probes actually went to Mars and sent back close-up photographs showing that the landscape was all arid deserts. You cannot see the Suez Canal or even the Nile River in Fig. 3.4, a photo taken of the earth from inside a spacecraft, so why should anyone expect to see canals on photos of Mars taken from tens of million miles away through our atmosphere? But I digress.

Nevertheless, I do not think that there is much room for doubt about what I saw through my telescope. No-one from Galileo on down has ever reported any different.

Ptolemy knew about the elongation problem. He proposed that Venus actually orbits the Earth in a circle going around a circle, known as an epicycle. Cunningly, the epicycle and the sun's motion are supposed to be synchronized in such a way that the angle SEV is never more than the greatest eastern or western elongation, about 47°.

In the scenario, shown in Fig. 1.16, in which Venus is closer to the Earth than the Sun, Venus can never have a gibbous phase when viewed from Earth. What happens if Venus is further away than the Sun? This is the scenario shown in Fig. 1.17.

Now, Venus is always gibbous as seen from the Earth.

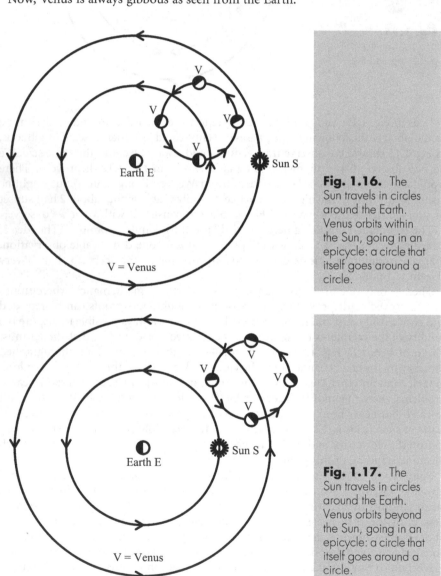

Fig. 1.16. The Sun travels in circles around the Earth. Venus orbits within the Sun, going in an epicycle: a circle that itself goes around a circle.

Fig. 1.17. The Sun travels in circles around the Earth. Venus orbits beyond the Sun, going in an epicycle: a circle that itself goes around a circle.

But I have seen for myself, and reported in Fig. 1.12, that Venus can show both crescent and gibbous phases from my backyard, which is most definitely on the same planet as you are, dear Reader.

Sorry, Ptolemy, you were just plain wrong. Does this mean you were a dummy? Mostly it means you never got to look through a telescope. I think you would have been very surprised.

When Can You Try This for Yourself?

Inferior conjunctions of Venus occur every 584 ± 4 days, which is every 19 months and a few days. Greatest elongations occur 71 ± 1 days beforehand for Eastern elongations and 71 ± 1 days afterward for Western elongations.

Table 1.1 gives dates of future events. If you read this in many years' time, you can use the above formulae to estimate suitable dates.

A rare event will occur on June 6 2012: a transit of Venus, when it will actually cross the Sun's disc. This will be visible from the British Isles just after sunrise, from the USA late in the day, and from Australia and New Zealand, except that it will start before the Western Australian sunrise. Please bear in mind that the best place to view this phenomenon is close to the International Date Line, so the date may be a day out for you.

The other dates carry a similar health warning – they may be a day out for you. Indeed, different sources sometimes vary by up to 2 days. That is one reason why I have cited two sources per date. The other reason is that tables do sometimes contain errors. The sources themselves often do not quote their own sources or say how they verified their claims. Since I wish they would be more careful about evidence, I can hardly look people in the eye if I am less careful.

Conclusion

I have shown how the size, phase, and timing of what I have observed about Venus are all consistent with the hypothesis that both the Venus and the Earth orbit the Sun in approximately circular orbits; that they obey Newton's laws of motion and of

Table 1.1. Venus' Inferior Conjunctions and Greatest Elongations

Greatest Eastern Elongation	Inferior Conjunction	Greatest Western Elongation
January 15 2009[2,8]	March 27 2009[2,8],	June 5 2009[2,8]
August 19 2010[2,8]	October 29 2010[2,8]	January 8 2011[8,9]
March 27 2012[9,10]	June 6 2012[10,11]	August 15 2012[9,10]

gravity; and that Venus orbits the Sun at a distance of approximately 0.73 AU taking approximately 228 days to complete one orbit.

I have also shown that the alternative hypothesis, favored in ancient times, that both the Venus and the Sun orbit the Earth, is utterly inconsistent with the telescopic evidence of the sizes and phases of Venus.

Finally, I have given some dates when you can check for yourself whether I am full of . . . you know what.

CHAPTER TWO

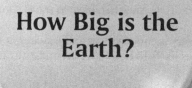

How Big is the Earth?

Ways to Measure the Earth's Size

The Flat Earth Society prospered in Britain in my youth. People traveled a lot less in those days. Since the whole country has about the land area of Colorado, and the populated part is smaller still, we never had to deal with time zones or climate zones. Hence you never had to face up to the Earth's roundness. Since we live on an island, we never drove into other countries before the Channel Tunnel opened in 1994.

On my first intercontinental journey, at the age of 22, it never occurred to me to reset my watch as the air stewardesses suggested; and to see broad daylight in Toronto at midnight jolted me into *experiencing* the roundness of the Earth. The effect was reinforced by the fact that Toronto is so much further South than England. It is at the same latitude as the South coast of France, a good thousand miles from here. The Sun was higher in the sky than I had ever seen it; and it got really dark early in the August evening.

Two years later I went to the United States as a postdoctoral researcher at Case Western Reserve University. Nobody in England had heard any Cleveland jokes, so I did not know any better. I turned down a job offer at Colorado State in beautiful Fort Collins to go there – can you believe it? No matter: I had a great time in Ohio, and met my wife, so to heck with Cleveland's detractors. While there I made a Greyhound bus trip to Phoenix, 2000 miles to the southwest of Cleveland. The day before leaving, I happened to notice the sunset time in Cleveland – 6 P.M. Three days later, with my watch unadjusted, I could see the Sun still quite high in the sky in Phoenix at 6 P.M. That impressed me: you cannot take a bus trip across time zones in England.

In essence, there are two effects of traveling large distances: the sunrise and the sunset times change, and things get higher or lower in the sky. You can use either one to quantify the size of the Earth.

J.D. Clark, *Measure Solar System Objects and Their Movements for Yourself!*, Patrick Moore's Practical Astronomy Series, DOI: 10.1007/978-0-387-89561-1_2, © Springer Science + Business Media, LLC 2009

It is a bit difficult to verify the number of miles you travel in an airliner: you have to take the airline's word for it that they are not diddling you over air miles. In a car it is different. You have an odometer, which you can calibrate against freeway mile-posts, and otherwise generally learn to trust.

Various challenges face the amateur measurer of the Earth's size. Most of us are unlikely to own specialist equipment. That particularly makes measuring latitude difficult. You can buy a cardboard model sextant for less than £50 ($100), which claims to be good to 5 min of arc. You then have to travel a long way North-South to detect an effect.

These days we all own very accurate clocks and watches. It therefore seemed to me that going east-to-west was likely to be a more accurate way to detect an effect of the Earth's roundness.

In either case, your next problem is to find a road that goes due east or due west for a long distance. In North America that's easier than in Europe. The only really long straight North-South road I found using Google Maps was in Argentina! The publisher's advance was never going to fund a ride along that. In Spain you can go more or less straight from North to South; and you can drive more or less East-West from Amsterdam to Warsaw. I considered all of these possibilities, but they seemed to be an expensive gamble in both time and cash when you cannot rely on the weather being clear. I even thought about going to Israel, where there is a long North-South road from the Lebanese border to the Red Sea with a high probability of clear skies. The trouble is that it does not go due North-South. Eilat is some way to the West of Kiryat Shemona.

What Johnny Did Next

I live close to the widest part of the island of Great Britain. There is actually a 25-min difference in sunset times between the East and West coasts at my latitude. Three months I waited for the weather to be clear right across the country; and I gave up. It just was not likely to happen before the publisher's deadline. Instead I settled for a 140-mile journey West, which would give me about a 12-min sunset time difference.

My wife stayed here. She used a cardboard solar projecting telescope, which we bought back in 1999 to view a total solar eclipse in Hungary. She was accompanied by our teenage daughter, who rather thought we were wasting her time.

Me, I headed off in the car with a borrowed 4-in. Newtonian with a solar filter and my 6-in. Newtonian.

The date we chose was going to be the last one before a short spell of fine weather broke and another Atlantic weather system blew in. "System" is of course a euphemism for horrible weather. It had already crossed Wales and was heading for England as I left.

It was the journey from Hell. The satellite navigator got lost – it obviously knew I had to get somewhere before sunset. If truth be told I think it knew where it was going, until it and I encountered a road closed for repair. It could not deal with the diversion. The other thing that made the going tough was that, in order to go as due west as possible, I had to go through every town center and avoid all the bypasses. So I caught a bad case of evening rush hour. The detour ruined my odometer reading.

So instead of returning home via the best roads, I had to retrace my route along some rather second rate roads chosen for their directness not their quality, in the dark. I will NEVER follow that route again!!!

That was a curtain raiser for what happened near sunset. I had carefully staked out a region of flat land, but the weather system came in faster than forecast. I was in somewhat hilly country, which is not the best place to time a sunset. In the end I stopped at a road junction just outside the village of Haughton in Staffordshire, where there was a good view of the sun, a reasonably flat view toward the horizon, and a grass verge where I could set up shop. This was eight miles short of my planned stop, but there was nothing I could do. The cloud was thickening by the mile.

Wife and daughter, meanwhile, had moved six miles North to get a better view of the horizon. They had no problems with cloud, but this did complicate my calculation slightly.

The cunning plan had been to record the times of first contact between Sun and horizon, the mid-point of the sunset, and the last view of the sinking Sun. This would give us three times to compare. There is a function in Microsoft Excel called "Now" which you can use to capture times. You press enter, then immediately copy the cell and go Paste Special -> Values to enter this time permanently in another cell. It all worked magnificently on the dry run. It worked a bit less well when I did it for real. It is a bit harder to do when you are holding a cell phone as someone talks you through the sunset back home. They saw the middle and final points of the sunset, and reported this to me. I could still just about see the Sun through cloud, so that felt a little bit weird.

Meanwhile, chez moi, the Sun became invisible before it reached the horizon. I was reduced to three ways to detect the sunset time. First, I had abandoned the borrowed telescope with a solar filter, and had aligned my motor driven 6-incher on the Sun. I had enough time to correct the alignment of the finder scope, and thereafter used that. So one measure of sunset time was when the cross-hairs of the finder scope reached the horizon. Fortunately I was able to make a pretty good guess which way North was, since I have an equatorial mount.

The second measure was that most of the sky was still clear. It was just the west that was cloudy. The clouds here were low; and there was some higher cloud, some of which was caused by aircraft vapor trails. At sunset, clouds often change color from white to pink to dark blue. They are pink right at sunset, because for a few seconds, the sun's rays are illuminating them from below. It happens to low clouds before high ones at any given location. I was able to watch all this, and make a second guess about the sunset time.

Third, my satellite navigator, which at this point was in the dog house, has a screen which switches from white to black at sunset. This gave me another measure.

Figure 2.1 shows the route I took. It was not very direct: English roads never are.

I later used three atlases to compare the distance I drove with the as-the-crow-flies distance between family and self. Two of them gave more or less the same distance; and one did not. Since our latitude is more or less halfway up England, my route was either going to be at the top of the Southern England page or at the bottom of the Northern England one. Either way, the projection was unlikely to be optimum.

The results are shown in Table 2.1.

If I used atlases to get the mileage, why did I bother to drive? For two reasons: first, to verify the map mileage, and second, the experiment was more powerful because

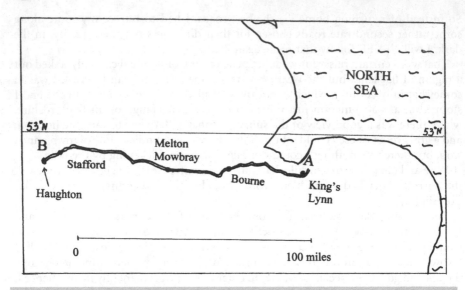

Fig. 2.1 The route taken.

Table 2.1. Constant-Latitude East-West Distance Between the Two Observers

Constant-Latitude Mileages from Atlases	Measured with Digital Calipers (mm)	Scale Measured with Digital Calipers	Miles/mm	Miles
Collins double atlas	194.3	65.25 mm = 60 km	0.571	111.02
Philips University atlas	196	50 mm = 50 km	0.621	121.79
Collins Europe atlas	178.76	30 miles = 47.17 mm	0.636	113.69
Average the two Collins results, reject the Philips one				112.35

we experienced the sunsets at different times 3 h drive apart. The whole idea was to see for ourselves how big the Earth is.

The car odometer read 131.9 miles for the journey back; and the satellite navigator read 133.6 miles. Google Maps said I drove 133 miles. You pays your money and you takes your choice. My route was about 18% longer than the constant-latitude distance. From the map in Fig. 2.1, that looks reasonable.

Figure 2.2 shows how to calculate the circumference of the circle of constant latitude along which the measurements were taken. The circumference of a circle at latitude L is $2\pi R \sin(L)$.

From the diary of events listed in Table 2.2, I concluded that the difference in observed sunset times was 11 ± 1 minutes. Since there are $24 \times 60 = 1{,}440$ min in a day, I traveled $(11 \pm 1)/1{,}440 = 0.00764 \pm 0.00069$ of the way around the Earth.

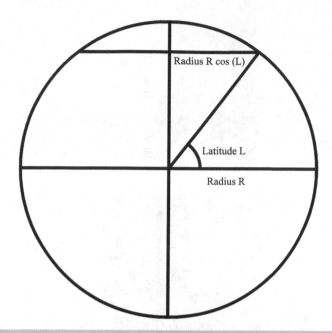

Fig. 2.2 Showing how to calculate the radius of the circle of constant latitude *L*. The radius is *R* sin(*L*). The circumference of this circle is therefore 2π*R* sin(*L*).

We know that the mean latitude of the observation points was 52.815° or 0.9218 radians. This can be checked against the angle the pole star makes to the ground in my backyard. I approximately do this every time I polar-align. Therefore

$$2\pi R \cos(0.9218) = \frac{112.35}{0.00764 \pm 0.00069};$$

$$R = \frac{112.35}{2\pi \cos(0.9218)(0.00764 \pm 0.00069)} \qquad (2.1)$$

$$= 3{,}873 \pm 352 \, \text{miles}.$$

This in turn gives a circumference of

$$2\pi(3{,}873 \pm 352 \, \text{miles}) = 24{,}336 \pm 2{,}212 \, \text{miles}. \qquad (2.2)$$

I am very impressed by the astronomical data in the online encyclopedia Wikipedia (http://en.wikipedia.org/wiki/Main_Page). Particularly during school vacations, Wikipedia articles are sometimes vandalized, so be a little careful. Converting the Wikipedia values (http://en.wikipedia.org/wiki/Earth) into miles gives a mean radius of 3,959 miles and a mean circumference of 24,881 miles.

Table 2.2 Diary of Events as Recorded

	King's Lynn	King's Lynn	King's Lynn	Haughton	Haughton	Haughton
First contact	08/05/2008		15:58:33	08/05/2008	20:39:58	First pink vapor trials
Midpoint		20:32	15:58:09	08/05/2008	20:43:19	Lowest clouds black
Last contact	08/05/2008	20:34	15:58:25	08/05/2008	20:44:47	Satellite navigator registers sunset
						Also noticed crosshairs of finder scope had reached horizon in last minute or so
				08/05/2008	20:50:08	All but highest clouds black

Conclusion

My mean values of the Earth's radius and circumference are a bit low (2.2%), but the published answer is well within my error range. My error actually represents just over 9%, with which I am not terribly satisfied.

What would I do differently? I think the first thing is that I would use sunrises. There is an inherent problem with sunsets, especially as far north as England. It starts to cool down in the evening before sunset, a phenomenon much less noticeable in my wife's native Seattle, which is a good 5° of latitude further South, even though the climates are somewhat similar. This in turn makes it no great surprise that clouds often form just before sunset.

There is no analogous phenomenon before sunrises, so I would use a sunrise to repeat the measurement. Sunrises here in June are around 4:30 A.M. Summer Time, which means (a) I would need a very willing accomplice and (b) there would not be much traffic to ruin my experiment.

CHAPTER THREE

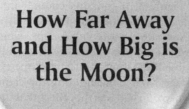

How Far Away and How Big is the Moon?

First Attempt

The story of this measurement is a classic example of the way in which the most important piece of equipment owned by an astronomer is his or her wit. This was not a planned measurement at all. I noticed something that surprised me, and followed it up.

One night in February 2007 I happened to notice that the Moon and Saturn were quite close together. If you know from a star chart roughly where Saturn is, you can easily identify it with the naked eye – not by its rings which require about 30x magnification – but by its creamy color. Like Mercury, Venus, Mars, and Jupiter, it is bright. These planets, unlike stars, do not twinkle.

Anyway, that night, I went to visit my parents. When I left, I looked up and saw the Moon again, but no Saturn. Puzzled, I got the telescope out and could then see Saturn right next to the Moon. This looked so cute it just had to be photographed with my brand new webcam.

The result was a disaster. Either the Moon was overexposed or Saturn was invisible. To make matters worse, they moved apart a lot faster than I expected. I went to try and find an old camera tripod to steady the webcam, which I had been hand-holding behind the eyepiece. By the time I got back, they were almost too far apart to see through the telescope even at minimum magnification.

Morale plummeted. Would I ever learn how to take astrophotographs?

Slowly, over the next day, it dawned on me that if I had been wider awake, I could have taken two photos showing both objects, and measured how far the Moon had moved; for it was the Moon that did most of the moving. It traverses the heavens once a month as it orbits our Earth. Saturn takes 29.46 years (353 and-a-bit months)

J.D. Clark, *Measure Solar System Objects and Their Movements for Yourself!*,
Patrick Moore's Practical Astronomy Series,
DOI: 10.1007/978-0-387-89561-3_3, © Springer Science + Business Media, LLC 2009

to orbit the sun and reappears in roughly the same place relative to the stars. (Saturn's motion is analyzed in Chap. 6.) It is therefore accurate to better than one part in 350 (99.7%) to assign all the relative motion to the Moon.

I looked...yes, I had enough pictures to do this.

Figure 3.1 shows what I had. Esthetically they are nonstarters. But they do contain usable information. They show the relative positions of the two celestial bodies as seen from my backyard. Figure 3.1b contains enough information to detect the orientation of the Moon. Figure 3.1a does not. Fortunately my bad technique saved me! The camera used, a webcam, shoots video clips, which of course consist of frames. Figure 3.2 shows what I got two frames earlier than Fig. 3.1a. The brightness of the image falls off big time if the camera and telescope eyepiece are not perfectly aligned. Because at this stage I was hand-holding the camera, the brightness was going up and down like a yo-yo. So bad was this effect that in two frames the pictures have gone from being exposed almost correctly for the Moon to being exposed about right for Saturn.

So I could use Fig. 3.2 to orient Fig. 3.1a. I printed Figs. 3.1a, b, and 3.2 onto transparent sheets of the type used for overhead projectors, overlaid them and tacked them together with adhesive tape. I then measured the "distance" between the two Saturns very carefully. Failing all else, you could do this with a ruler, but I actually used a set of calipers, Fig. 3.3, which are accurate to 0.01 mm.

I did my sums, and got a value for the Moon's diameter about 25% higher than the commonly published value of 3,476 km or 2,160 miles.

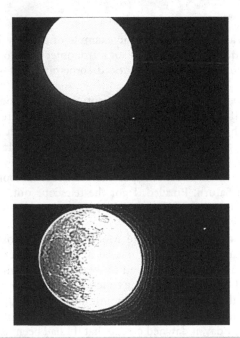

Fig. 3.1. Two terrible photos, but you can see that the Moon and the Saturn have moved apart in 38 min.

Fig. 3.2. Two frames earlier than Fig. 3.1a, this frame shows some features to enable me to orient the Moon in the picture.

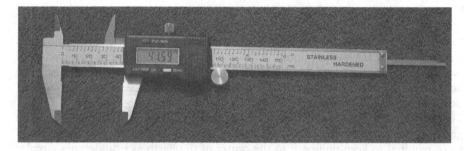

Fig. 3.3. Digital calipers.

Morale was restored. I had not wasted my time. The following month, weather permitting, I would have a go at repeating the experiment in a more planned fashion.

By the way, I take most of my published astronomical data from the Cambridge Planetary Handbook.[12] Another good source of scientific data is Kaye & Laby's Tables of Physical and Chemical Constants.[13] Google is also a good way to find data, and I have mentioned Wikipedia.

Calculations Using no more than High School Mathematics

Measuring the movement of the Moon against the heavens eliminates the effect of the Earth's rotation. What I really wanted was a "fixed" star to define the position of the heavens. The difficulty is that the Moon is so much brighter than most stars that you cannot see them in the telescope at the same time. This is because, unless your telescope is much better than mine, stray moonlight reflects off the interior walls of the telescope and eyepiece causing a "whiteout." Stars are not bright enough to outshine the stray moonlight. Planets work better because they are bright. Saturn moves sufficiently slowly through the heavens to benchmark lunar motion.

From such photos as Fig. 3.1, you can work out the speed of the Moon's motion in lunar diameters per hour. By using some general knowledge and a bit of cunning, I separately worked out the orbital speed of the moon in miles per hour. If I know the speed in miles per hour and diameters per hour, then it is dead easy to work out the numbers of miles per diameter.

Hold on! The number of miles per diameter is the diameter. That is what I wanted in the first place.

At this point I have a terrible confession to make. I always do calculations in metric units, because that is what I was brought up with. I will then happily convert kilometers to miles, because like most native English speakers, I think in miles. It does not matter one bit what units you use. They are all as good as one another. In my experience, you make fewer errors if you stick to the units you know.

How did I get the Moon's speed in miles per hour? By using a simplified version of what Isaac Newton did.

His great contribution to our knowledge of gravity was to prove that the Moon is kept in orbit by the same force that pulls terrestrial objects to the ground.

All the formulae relating to gravity and orbital motion that I have used can be found in textbooks. I did not invent any of them; and since my ancestors in the 1600s were not exactly big shots, I doubt if I would have had the necessary education to participate in science even had I been alive. In 34 years, I have not found one I like better than the one by Tony French. I once met him briefly. He is very friendly and down to earth; and it shows in his book.[14]

According to Newton, a circular orbit around the Earth is stable when the gravitational force on the satellite is exactly enough to provide the "centripetal" force required to keep pulling it to the center of the Earth as it undergoes circular motion.

His law of gravity states that the force F attracting two point masses

$$F = \frac{Gm_1 m_2}{r^2},\qquad(3.1)$$

where m_1 is the mass of the first point mass; m_2 is the mass of the second point mass; r is the distance between the two point masses; and G is a constant called the universal gravitational constant, often called "Big G."

Big G is the number that tells you how strong gravity is. Not how strong it is locally, but how strong it is throughout the Solar System. The bigger it is, the stronger gravity is. Whether it has always had the value it has today; and whether that value is the same everywhere, is still not known. Another way of describing Newton's great achievement is to say that he found that throughout the then known Solar System, Big G has the same value.

Equation (3.1) applies to point masses but the Earth and Moon are, to an excellent approximation, spheres rather than point masses. What set Newton apart from his contemporaries was that he found a way of dealing with this difficulty, and they did not. He proved the remarkable fact that spheres gravitationally attract exactly as if they were point masses located at their own centers. Newton's proof does not require more than an elementary knowledge of calculus, but it is a little more difficult than most of the mathematics presented here. It is such a fundamental result to planetary science that I have included Newton's proof in modern mathematical language later in this chapter. Less mathematically confident readers may skip that section.

For now, please let me treat the Earth's gravity as if it were generated by a point mass at the center. Then I could calculate the weight – the force due to gravity – of an object at ground level from (3.1) if I knew the relevant quantities.

Actually, weights are easy to measure – every household has the requisite equipment. It would be more useful to use that knowledge to back-calculate some other quantities in (3.1), which we cannot measure on the kitchen scales.

In order to apply it to a body of mass m_2 on the Earth's surface, acted on by the Earth's gravity, I am going to rewrite (3.1) as

$$F = \frac{GM_E m_2}{r_E^2} = \frac{GM_E}{r_E^2} m_2 = gm_2, \tag{3.2}$$

where the subscript E means "Earth" and body 2 is on the Earth's surface, and

$$g = \frac{GM_E}{r_E^2} \tag{3.3}$$

is called the acceleration due to gravity. It is also known is "little g." We know that it is acceleration because according to Newton's second law of motion,

$$\text{Force } F = \text{acceleration} \times \text{mass} = am. \tag{3.4}$$

If the mass is m_2, comparison of (3.2) with (3.4) gives

$$F = gm_2 = am_2. \tag{3.5}$$

Therefore little g, like a, is an acceleration. Its value is easily measured, and is well known to be

$$g = 9.81 \text{ m s}^{-2}. \tag{3.6}$$

This value is good in London and Washington, DC. It is a little higher at the poles and lower at the equator (Fig. 3.4).

We also need to know the Earth's radius, r_E. We noted the published value in Chap. 2 to be

$$r_E = 3{,}959 \text{ miles}. \tag{3.7}$$

Why are we doing all these mathematical gymnastics? In the short term, our objective is to avoid having to measure either Big G or M_E. G is hard to measure because the gravitational attraction between lumps of stuff in a lab is incredibly weak. To know the Earth's mass, we need to know all sorts of information about how dense it is 1,536¾ miles below the surface, which we would rather not have to worry about or we would be here forever. Indeed, the only realistic way to measure the Earth's mass is to measure G in a laboratory and use (3.2).

I am now going to rearrange (3.3) and substitute the result of (3.7) for the Earth's radius r_E. I then find that

$$GM_E = gr_E^2 = 9.81 \times (6.373 \times 10^6)^2 = 3.984 \times 10^{14} \text{ m}^3 \text{ s}^{-2}. \tag{3.8}$$

Fig. 3.4. Most adults from developed countries have now traveled enough to know that the Earth is round, even if we did not experience the wonderful views seen by the Apollo astronauts. Notice how tiny all the nonspherical effects are. The continents do not look higher than the oceans. They could almost have been painted on. There are mountainous regions not covered by cloud in Antarctica and Saudi Arabia. These do not look raised. The Earth does not look any flatter at the South Pole.

This may have been a struggle, but we have used our brains here to save our brawn, because we have worked out the value of Big G times M_E without having to know either of these difficult-to-measure quantities.

This completes the extraction of information from ground-level gravity.

I am now going to go back up to the Moon, so to speak, and use my ground-level information to help me work out what is happening with the Moon's orbit.

In order to do this, I am going to make a simplification. As Kepler taught us back in the early 1600s, orbits are really ellipses not circles.

A circle is a special case of an ellipse, where the major and minor axes are equal (Fig. 3.5). For those interested, the most readable book I have ever found about ellipses is the one by Thomas,[15] which I stole from my wife. One measure of the deviation of an ellipse from circularity is its *eccentricity*, defined by

$$\varepsilon = \sqrt{1 - \left(\frac{\text{minor axis}}{\text{major axis}}\right)^2}.$$ (3.9)

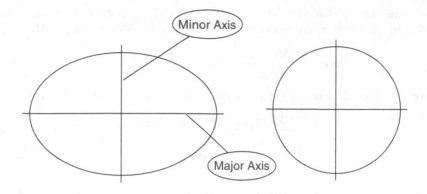

Fig. 3.5. A circle can be thought of as an ellipse whose major and minor axes are equal.

When the minor axis and major axis are equal, $\varepsilon = 0$. For the Moon's orbit, $\varepsilon = 0.0549$.[2] Back substituting into (3.9) gives

$$\varepsilon = 0.0549 = \sqrt{1 - \left(\frac{\text{minor axis}}{\text{major axis}}\right)^2},$$

$$\therefore \frac{\text{minor axis}}{\text{major axis}} = 0.998. \tag{3.10}$$

In other words, the major and minor axes of the Moon's orbits are the same to within 0.2%. We therefore treat the Moon's orbit as circular.

The centripetal force pulling a body into circular motion is given by

$$F_{\text{CIRC}} = \frac{m_{\text{moon}} v_{\text{moon}}^2}{r_{\text{orbit}}}, \tag{3.11}$$

Where m_{moon} is the mass of the Moon, v_{Moon} is the velocity of the Moon as it travels in its orbit, and r_{orbit} is the distance from the earth's center to the Moon's center. French's book gives a noncalculus derivation of this formula, which takes a page. (I have another book which uses calculus and gets there in three lines. That is why it is worth learning calculus.)

The time taken for a complete orbit T_{orbit} is the time taken to complete one circumference. Since speed × time = distance,

$$v_{\text{moon}} T_{\text{orbit}} = 2\pi r_{\text{orbit}},$$

$$\therefore v_{\text{moon}} = \frac{2\pi r_{\text{orbit}}}{T_{\text{orbit}}}, \tag{3.12}$$

since circumference = 2π × radius. Substituting (3.12) for v_{moon} into (3.11) gives

$$F_{\text{CIRC}} = \frac{m_{\text{moon}} 4\pi^2 r_{\text{orbit}}^2}{r_{\text{orbit}} T_{\text{orbit}}^2} = \frac{4\pi^2 m_{\text{moon}} r_{\text{orbit}}}{T_{\text{orbit}}^2}. \tag{3.13}$$

As we have stated, what we need to do is to equate F_{CIRC} to the gravitational force exerted by the Earth on the Moon, i.e., the weight of the Moon. Using (3.1),

$$F_{\mathrm{GRAV}} = \frac{GM_E m_{\mathrm{moon}}}{r_{\mathrm{orbit}}^2} = F_{\mathrm{CIRC}} = \frac{4\pi^2 m_{\mathrm{moon}} r_{\mathrm{orbit}}}{T_{\mathrm{orbit}}^2}. \qquad 3.14$$

We can cancel the mass of the Moon out of (3.14) and do a little bit of algebra:

$$\frac{GM_E m_{\mathrm{moon}}}{r_{\mathrm{orbit}}^2} = \frac{4\pi^2 m_{\mathrm{moon}} r_{\mathrm{orbit}}}{T_{\mathrm{orbit}}^2}$$

$$\therefore \frac{GM_E}{r_{\mathrm{orbit}}^2} = \frac{4\pi^2 r_{\mathrm{orbit}}}{T_{\mathrm{orbit}}^2}$$

$$\therefore GM_E T_{\mathrm{orbit}}^2 = 4\pi^2 r_{\mathrm{orbit}}^3 \qquad (3.15)$$

$$\therefore \frac{GM_E T_{\mathrm{orbit}}^2}{4\pi^2} = r_{\mathrm{orbit}}^3$$

$$\therefore r_{\mathrm{orbit}} = \sqrt[3]{\frac{GM_E T_{\mathrm{orbit}}^2}{4\pi^2}}$$

We already know the value of GM_E from (3.8). Therefore if we know T_{orbit}, we know r_{orbit}. The time taken for one lunar orbit is of course 1 month.

Well, almost. There is a slight subtlety. Figure 3.6 shows that the orbital period is slightly less than the time from New Moon to New Moon (or Full Moon to Full Moon). Due more to good luck than good judgment, I was treated to a dramatic confirmation of this when I repeated the experiment the following month. I was blessed with clear weather when the Moon next passed Saturn. On 3 February 2007, when Figs. 3.1 and 3.2 were shot, the Moon was about 17 h past full. In the early hours of March 2 2007, when the next close encounter with Saturn occurred, it was not quite full. This was very dramatically proved around midnight of 3–4 March 2007, two nights later, because there was a total lunar eclipse, which means that the Moon must have been full then, not on 2 March. (All times are Greenwich Mean Times.)

An additional subtlety, which we have neglected, is that the Earth and the Moon actually orbit around their common center of mass.

Using Saturn as a benchmark, I found T_{orbit} to be about 27.1 days, from 3 min past midnight on 3 February to 02:31 on 2 March. The published value is 27.322 days or 2.3606×10^6 s; the discrepancy between the measured value and the published value is about 0.8%.

We are now in a position to work out r_{orbit}. From (3.8) and (3.15), using the measured lunar orbital time of 27.1 days,

$$r_{\mathrm{orbit}} = \sqrt[3]{\frac{GM_E T_{\mathrm{orbit}}^2}{4\pi^2}} = \sqrt[3]{\frac{3.984 \times 10^{14}\ \mathrm{m^3\ s^{-2}} \times (27.1 \times 24 \times 3{,}600\ \mathrm{s})^2}{4\pi^2}}$$

$$= \sqrt[3]{\frac{2.1842 \times 10^{27}\ \mathrm{m^3}}{39.478}} = 3.810 \times 10^8\ \mathrm{m}. \qquad (3.16)$$

This is my distance to the Moon. As a sanity check, (3.16) can again be compared against published values. The half-major axis is published[16] to be 3.844×10^8 m.

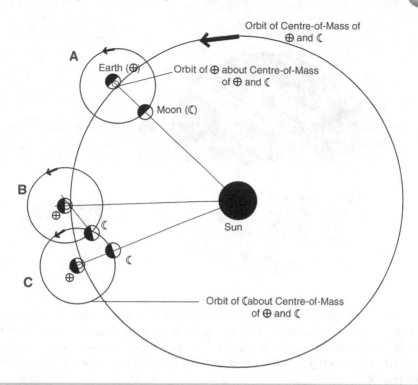

Fig. 3.6. Exactly what is the time taken to complete one orbit? New Moons occur at A and C. This is because the Moon is in the same direction as the Sun when viewed from the Earth. In the meantime, the Earth has moved about one-twelfth of the way around the Sun. (This movement is exaggerated in the figure for clarity.) By the time it is at C, the Moon has completed more then one orbit since it was at A. It completed this orbit at B. Therefore the orbital period is slightly less than the New-Moon-to-New-Moon time.

From (3.10), this implies a half-minor axis of 3.836×10^8 m. The discrepancy between the average of these and my result, (3.16), is about 0.8%.

Thus, I have a very good answer to the first of the two questions posed by the chapter title. Let us move on to the second question.

Advanced Topic: A Sphere Attracts Like a Point Mass

In this section, I carry out my threat to give Newton's proof that a sphere gravitationally attracts like a point mass. You will need some elementary calculus to follow this, but nothing advanced. I am using the lowest-tech proof I know of. If you do not want to read this now, you can skip to the next section without loss of continuity.

Fig. 3.7. Imagine a thin sphere of material within our planet, centered on the planet's center.

You break the sphere down into manageable chunks and analyze their gravitational attraction. First, suppose that the planet is perfectly spherical: this is out by about 0.33% for the Earth,[17] so it is a good enough approximation. Now imagine a thin sphere within our planet like the one in Fig. 3.7, which is centered on the center of the planet.

I actually want you to imagine that this sphere-within-a-sphere is infinitesimally thin. Suppose that our planet has radius R_P, and that the internal sphere has radius R.

What we are going to do is to work out the gravitational attraction of this sphere on a point mass m outside the planet. I agree that "point mass" is an oxymoron, but we will see that this does not matter. Figure 3.8 shows how the sphere-within-a-sphere is further broken down into infinitesimally thin rings.

Let each element of the ring shown in Fig. 3.8 be at a distance l from the point mass m. Suppose further that our sphere has uniform density ρ. By elementary geometry, the volume of the ring is

$$V_{\text{ring}} = (\text{circumference})(\text{height})(\text{thickness}) = (2\pi R \sin(\theta))(R\,d\theta)(dR). \quad (3.17)$$

Hence the mass of the ring is

$$m_{\text{ring}} = (2\pi R \sin(\theta))(R\,d\theta)(dR)\rho = 2\pi\rho R^2 \sin(\theta)d\theta\,dR. \quad (3.18)$$

It is easier to re-express this mass as a fraction of the mass of the shell, which I will call dM.

$$dM = 4\pi\rho R^2 dR, \quad (3.19)$$

since the surface area of the shell is $4\pi R^2$, and its thickness is dR. Now,

$$m_{\text{ring}} = 2\pi\rho R^2 \sin(\theta)d\theta\,dR = \frac{1}{2}dM \sin(\theta)d\theta. \quad (3.20)$$

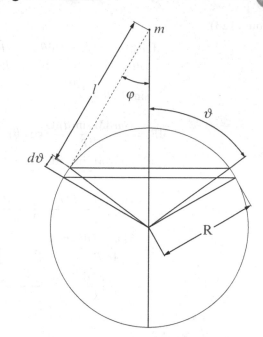

Fig. 3.8. Our internal shell of matter acts gravitationally on a point mass m at a distance r from the center of the planet. For analysis, the shell is to be further divided into infinitesimally thin rings, each of whose centers is coincident with the line from the planet's center to the point mass m. Each of the also rings lies in a plane normal to the line from the planet's center to the point mass m.

The force this ring exerts on the point mass is therefore, by Newton's law of gravity (1.16),

$$dF = -\frac{Gm m_{\text{ring}}}{l^2} = -\frac{Gm\,dM\,\sin(\theta)d\theta}{2l^2}. \tag{3.21}$$

The minus sign indicates that m is being pulled toward the planet.

There is actually an error in (3.21). Some of the forces I counted in (3.21) cancel out. All those components that act perpendicular to the line from the planet's center to the point mass m cancel. The parts that do not cancel are those from each point on the ring that act parallel to the line from the planet's center to m. Equation (3.21) should therefore read

$$dF = \frac{Gm\,dM\,\sin(\theta)d\theta}{2l^2}\cos(\varphi). \tag{3.22}$$

To get the gravitational force due to the whole thin shell shown in Fig. 3.7, we need to integrate (3.22) over all values of θ. This is not quite as easy as it looks because l and φ also depend on θ.

I need to do a little bit of geometry to simplify (3.22). I use the cosine rule (A.44) to do this. I need to apply this rule twice. The first time I get

$$\cos(\theta) = \frac{r^2 + R^2 - l^2}{2rR}; \tag{3.23}$$

and the second time I get

$$\cos(\phi) = \frac{r^2 + l^2 - R^2}{2rl}. \tag{3.24}$$

From (3.23)

$$\frac{d\cos(\theta)}{dl} = \frac{d\cos(\theta)}{d\theta}\frac{d\theta}{dl} = \frac{-2l}{2rR} = \frac{-l}{rR}$$

$$\therefore \sin(\theta)d\theta = \frac{l\,dl}{rR}, \tag{3.25}$$

$$dF = \frac{Gm\,dM\,\sin(\theta)d\theta}{2l^2 Rr}\cos(\theta)$$

$$= \frac{Gm\,dM\,\sin(\theta)d\theta}{2l^2 Rr}\left[\frac{r^2 + l^2 - R^2}{2rl}\right]$$

$$= \frac{Gm\,dM\,l\,dl}{2l^2 Rr}\left[\frac{r^2 + l^2 - R^2}{2rl}\right] \tag{3.26}$$

$$= \frac{Gm\,dM\,dl}{4r^2 R}\left[\frac{r^2 + l^2 - R^2}{l^2}\right]$$

$$= \frac{Gm\,dM\,dl}{4r^2 R}\left[1 + \frac{r^2 - R^2}{l^2}\right].$$

$$F = \int_{r-R}^{r+R} dl\,(dF)$$

$$= \int_{r-R}^{r+R} dl\left[\frac{Gm\,dM}{4r^2 R}\right]\left[1 + \frac{r^2 - R^2}{l^2}\right]$$

$$= \left[\frac{Gm\,dM}{4r^2 R}\right]\int_{r-R}^{r+R} dl\left[1 + \frac{r^2 - R^2}{l^2}\right]$$

$$= \frac{Gm\,dM}{4r^2 R}\left\{\int_{r-R}^{r+R} dl + \int_{r-R}^{r+R} dl\left[\frac{r^2 - R^2}{l^2}\right]\right\}$$

$$= \frac{Gm\,dM}{4r^2 R}\left\{[l]_{r-R}^{r+R} + \left[-\frac{(r^2 - R^2)}{l}\right]_{r-R}^{r+R}\right\} \tag{3.27}$$

$$= \frac{Gm\,dM}{4r^2 R}\left\{\{r + R - r + R\} + \left[-\frac{(r+R)(r-R)}{l}\right]_{r-R}^{r+R}\right\}$$

$$= \frac{Gm\,dM}{4r^2 R}\left\{2R - \left[\frac{(r+R)(r-R)}{(r+R)} - \frac{(r+R)(r-R)}{(r-R)}\right]\right\}$$

$$= \frac{Gm\,dM}{4r^2 R}\left\{2R - [(r-R) - (r+R)]\right\}$$

$$= \frac{Gm\,dM}{4r^2 R}\left\{2R - [-2R]\right\}$$

$$= \frac{Gm\,dM}{r^2}.$$

So, after a few pages of calculation, we see that the gravitational attraction of the shell is the same as that of a point mass dM acting at its center. It does not depend on the value of R. Isn't that amazing?

We still do not have the gravitational attraction of the whole planet on m. Now, I want to point out something else remarkable. If we imagine a second concentric shell, with a different radius, and with a density $f\rho$, where f is a factor which may or may not equal one, we have just proved that it too will attract m as if it were a point mass at the planet's center. Therefore, the shells do not have to have the same density. Each shell has to have a constant density, but the shells themselves can vary in density. This is an important result, because we hear that the earth has an iron core, and iron is much denser than the rocks and soil we see at ground level.

Strictly speaking, F is still an infinitesimal force, which we need to integrate from the planet's center to its surface to get the total force. The answer is

$$F_P = \frac{Gm}{r^2} \int_0^{R_P} dM$$

$$= \frac{Gm}{r^2} \int_0^{R_P} 4\pi R^2 \rho(R) dR \qquad (3.28)$$

$$= \frac{GmM_P}{r^2},$$

where F_P is the gravitational force due to the whole planet, and M_P is the mass of the planet, obtained by integrating over the shells at each radius R, each of which has volume $4\pi R^2 dR$ and density $\rho(R)$. $\rho(R)$ can depend or R but not on any angle.

The small mass m does not have to be a point mass either. It too can have spherical symmetry about its center; and it can even be large. The theorem we have just used still holds, so long as both bodies are spherically symmetric about their centers.

Ok, you object, what about the oceans? They are not as dense as the seabed or it would float! Does not that ruin the spherical symmetry of the Earth? Good question. I think the answer is that they are, on a global scale, very shallow. Their average depth is only a couple of miles.[18]

Second Experiment

To get a better estimate of the size of the Moon, I carried out my threat to repeat the experiment of 3 February.

There was actually an "occultation" of Saturn on 2 March 2007. This means that, from my location, the Moon passed right in front of Saturn. This did not make too much difference to the quality of my measurement, but it was fun to watch.

The first thing I did different was to take lots of photos. Remembering my earlier experience, I exposed the photos in pairs, one to catch Saturn, one for the features on the Moon. This turned out not to be necessary, but this did not matter: I was not taking any chances. As Fig. 3.9 shows, I was able to use a pair of mountains along the terminator to orient the Moon.

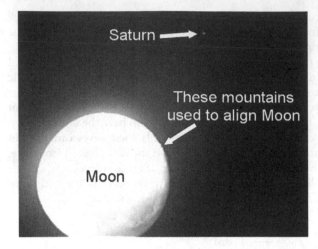

Saturn ➡️ ·

These mountains used to align Moon ↙

Moon

Fig. 3.9. Photo of the Moon and Saturn taken on 2nd March 2007 at 01:22 GMT. The Moon was not quite full. Mountains could be seen along the "terminator", the line of sunrise. Two of these were big enough to enable accurate orientation of the Moon in the pictures.

The sequence of pictures I shot is shown in Fig. 3.10.

I could have used the same method as before: to overlay transparent images to measure the movement of Saturn relative to the Moon. This would have been cumbersome. Necessity became the mother of invention, and I tried a new idea.

CAD (Computer Aided Drafting) software is normally used for producing engineering drawings. I used a package called TurboCAD^{TM} (http://www.turbocad. com). I used the professional version from work, but I am sure the entry level version would have done. Indeed, there is plenty of free CAD software available. I have tried Alibre Express^{TM} (http://www.alibre.com/xpress/?source = LTFRB2006). It can do what is required here, and is free.

I imported the photos from Fig. 3.10 into TurboCAD, and then used the features common to CAD systems to overlay circles and lines. Three points uniquely define a circle. Most CAD systems enable you to draw the circle that passes through three points. By choosing points on the edge of the Moon (but not on the terminator), I produced circles which coincided with the Moon. The CAD software can start a line from the exact center of this circle. This I did, and ended the line at the middle of Saturn, as accurately as I could estimate it. The zoom facility in the CAD system enabled me to place the line in the center of the brightest pixel in the middle of the image of Saturn. The CAD software enables you to measure the length of this line, and the diameter of the circle very accurately. (I chose three significant figures.) I then put another line from the center of my circle to the top of the benchmark mountain, which I was using to orient the Moon. Again, the CAD software told me the angle this line made with the line to Saturn to my chosen accuracy, 0.01°. On each photo in the sequence, I repeated this procedure three times, to give me a handle on the level of error I was making when guessing positions of things.

An example of this process is shown in Fig. 3.11.

Now I had a sequence of angles and distances from the center of the Moon to Saturn. I turned these into distances in x and y coordinates, using the line from the Moon's center to the reference mountain as my x-axis. These are plotted in Fig. 3.12.

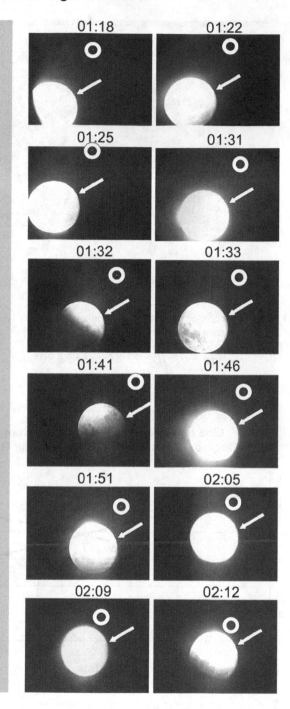

Fig. 3.10.
Sequence of pictures of the Moon and the Saturn. The white arrows indicate the position of the orientation feature.

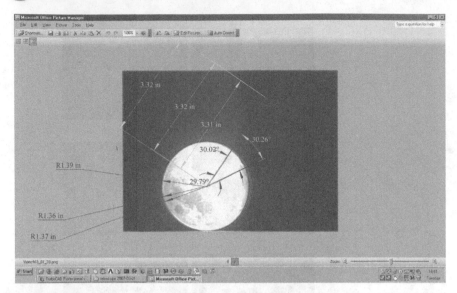

Fig. 3.11. Example of measurements made using the CAD software. It can be seen that there is a little bit of scatter in the data. In this case I found Moon diameters of 1.40, 1.42, and 1.43 in.; Moon-to-Saturn distances of 3.77, 3.77, and 3.79 in., and angles relative to the reference mountain of 27.10°, 27.46°, and 27.80°.

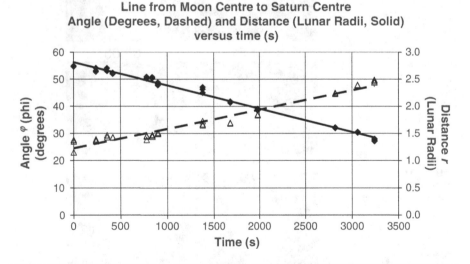

Fig. 3.12. Graph showing the position of the Saturn relative to the Moon at various times. The distance units are Lunar radii, as measured per the process shown in Fig. 3.11.

Extracting the Moon's Speed in Radii per Second Using only High School Mathematics

Unfortunately, the information presented in Fig. 3.12 is not in a very useful form. What we really need is to find the direction in which the Moon and Saturn travel toward each other. The information is there: it needs to be extracted.

I will do this in a way that minimizes the amount of mathematics needed. For the first step, a little trigonometry is unavoidable. The positions of Saturn relative to the center the Moon in Fig. 3.12 have to be re-expressed as x and y coordinates. The x-axis will be the line from the Moon's center to my reference mountain. Of course, it is best to do the analysis in a spreadsheet package like Microsoft Excel^{TM} or OpenOffice.org Calc^{TM}. The formulae to convert each data point are

$$x_i = r_i \cos\left(\frac{\pi\varphi_i}{180}\right);$$
$$y_i = r_i \sin\left(\frac{\pi\varphi_i}{180}\right),$$

(3.29)

where φ is the Greek letter *phi* and the subscript i refers to the ith measurement. The factor $\pi/180$ is to convert the units of the angle from degrees to radians. Most spreadsheets require this. Everyone knows what degrees are: they are what you get when you divide a complete circle into 360 equal angles. Radians are merely another unit. There are 2π radians in a circle. This is not a whole number, but who cares? Anyway, the conversion factor is

$$1° = \frac{2\pi}{360} \text{ radians} = \frac{\pi}{180} \text{ radians.}$$

(3.30)

Next, we get Microsoft Excel 2003 and our CAD system to do the mathematics for us.

In Fig. 3.13, I have added a "trendline" to those points which were measured between 900 and 3,060 s after the start of the experiment. I rejected the others because I could not reproducibly guess where the Moon's surface was on the photos. To add a trendline in Excel 2003, click with your secondary mouse button (the one under your third finger) with the cursor over blank white area, and it will appear in the menu as an option.

Incidentally, I did not say "right click" because I am in the left-handed 10% of the population, and have my mouse buttons set the opposite way to most people.

I have finished with Excel for now. I made a screen dump of the graph in Fig. 3.13, and saved it in a form that I can download into TurboCAD. I actually installed one of the many free PDF creators that you can get, and "printed" the graph to a pdf file, from which I copied and pasted the graph into TurboCAD. You can also save the graph as a jpeg or a png using one of the picture editor programs, and then insert it into the CAD package.

First, I stretched the graph vertically until the x and y scales were the same. Having one lunar radius equal to 2.097 in. one way and 2.109 in. the other way is close enough for our purposes. The average is one lunar radius = 2.103 in. The inches are relative to the CAD software's internal scale.

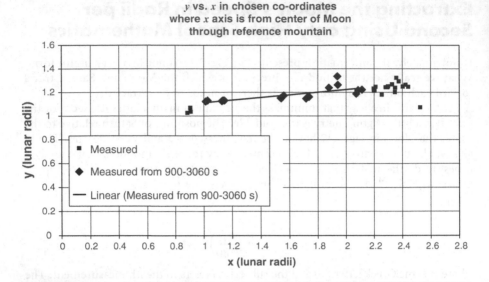

Fig. 3.13. The data shown in Fig. 3.12 re-expressed as distance from the Moon to the Saturn in *xy* coordinates.

Next, I drew a straight line over the trendline fitted by Excel, and extended it to beyond the left-hand end of the graph. It is shown in Fig. 3.14. I drew another line along the grid line at $y = 1$ lunar radius, which intersects the line I just drew. The CAD software then tells me that the angle between these two lines is 6.58°.

Next comes the slightly clever bit. All the relative motion of the Moon is along this line at 6.58° to the horizontal. Between 900 and 3,060 s, the relative motion is from one end of the trendline to the other. This is a distance of 2.33 in. Since we know that one lunar radius is 2.103 in., we can use the well-known formula that speed is distance divided by time to give

$$\text{Speed} = \frac{(2.33/2.103)}{(3,060 - 900)} = \frac{1.1079}{2,160} = 5.13 \times 10^{-4} \text{ lunar radii s}^{-1}, \qquad (3.31)$$

where in the usual way, s^{-1} means "per second."

Extracting the Moon's Speed in Radii per Second Using Some College Mathematics

If you know a little more mathematics, you can obtain the result of (3.31) with a little more knowledge of what you have done.

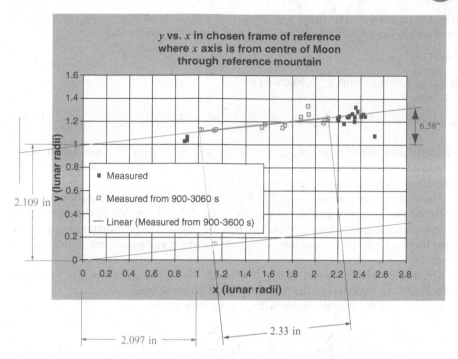

Fig. 3.14. The graph of Fig. 3.13 after processing it in TurboCAD.

First, you can use linear regression formulae for yourself to fit a line through the data points in Fig. 3.14. Linear regression is described and derived in the updated edition of Spiegel's book on statistics[19]; and a justification for using least-squares fit is given in the Statistical Appendix.

A straight line is fitted through n points (x_i, y_i), namely

$$y = a + bx, \tag{3.32}$$

where a and b are fitted constants.

I will now show you how values for a and b in (3.32) can be worked out. I am going to use the principle of least squares. First, the fitted value y_f of y at $x = x_i$ is

$$y_f = a + bx_i. \tag{3.33}$$

Therefore the difference between the fitted and measured values of y at the ith data point is

$$y_f - y_i = a + bx_i - y_i. \tag{3.34}$$

Minimizing these differences is desirable. The trouble is that some will be positive and some negative. This messes up the minimization exercise. Let us use a bit of cunning: the quickest way to get a set of numbers that are all positive, but represent the differences between fitted and measured values, is simply to square the little sons-of-bitches. When you do this, you find that (3.34) gives you

$$(y_f - y_i)^2 = (a + bx_i - y_i)^2. \tag{3.35}$$

If we add all the terms in (3.35), we get

$$\sum_{i=1}^{n} (y_f - y_i)^2 = \sum_{i=1}^{n} (a + bx_i - y_i)^2. \tag{3.36}$$

The left-hand side of (3.36) is the sum of the squares of the differences between the fitted and the measured values of y at each of the i points. Minimizing the sum of the squares of the differences gives a good fit to (3.34).

There is one assumption which I have made but not told you about. I have assumed that the error in the values of x is much less than that in the values of y. In this case, there is no reason to believe that we know the x_i any better or any worse than they y_i. I ask you to hold this thought for a while, and permit me to deal with it later.

Meanwhile, please do not worry too much about the details of the following calculation: an overview of the principles is all you will need.

Equation (3.36) is a minimum with respect to a when

$$\frac{\partial}{\partial a} \sum_{i=1}^{n} (y_f - y_i)^2 = \frac{\partial}{\partial a} \sum_{i=1}^{n} (a + bx_i - y_i)^2$$

$$= \sum_{i=1}^{n} 2(a + bx_i - y_i) = 0. \tag{3.37}$$

Similarly, (3.36) is minimum with respect to b when

$$\frac{\partial}{\partial b} \sum_{i=1}^{n} (y_f - y_i)^2 = \frac{\partial}{\partial b} \sum_{i=1}^{n} (a + bx_i - y_i)^2$$

$$= \sum_{i=1}^{n} 2x_i(a + bx_i - y_i) = 0. \tag{3.38}$$

Many textbooks, such as Spiegel's, lay out the mathematical manipulation to solve (3.37) and (3.38). The solutions for a and b are

$$a = \frac{\left[\left(\sum_{i=1}^{n} x_i^2\right)\left(\sum_{i=1}^{n} y_i\right)\right] - \left[\left(\sum_{i=1}^{n} x_i y_i\right)\left(\sum_{i=1}^{n} x_i\right)\right]}{\left[n\left(\sum_{i=1}^{n} x_i^2\right)\right] - \left[\left(\sum_{i=1}^{n} x_i\right)^2\right]};$$

$$b = \frac{\left[n\left(\sum_{i=1}^{n} x_i y_i\right)\right] - \left[\left(\sum_{i=1}^{n} y_i\right)\left(\sum_{i=1}^{n} x_i\right)\right]}{\left[n\left(\sum_{i=1}^{n} x_i^2\right)\right] - \left[\left(\sum_{i=1}^{n} x_i\right)^2\right]}. \tag{3.39}$$

The key point is that the fitted line, (3.33), only allows for errors in the y_i, not the x_i. it is possible to derive a line-fitting formula that treats errors in the y_i and x_i equally. The conceptual method is the same, but the mathematical manipulation is lengthy. A quick and dirty way to estimate the effect is to see if we obtain the same answer when the axes are reversed.

You can either do this using (3.39) with x and y the original way round, then with them reversed; or you can repeat the CAD method described in the previous section.

Wait a minute...reversing x and y is not bright. It is smarter to rotate the coordinates by 90° to avoid making a mirror-image coordinate system.

I rotated the coordinates clockwise by 90°. The usual convention in coordinate geometry is that clockwise angles are negative, and anticlockwise ones are positive. I therefore subtracted 90° from every angle I had measured. My new y-axis points to the reference mountain, e.g., in Fig. 3.9 and my new x-axis points downward at right angles to it.

In this coordinate system, I repeated the method described in the previous section, and obtained the result as shown in Fig. 3.15.

From Fig. 3.15 I find a travel of 1.095 lunar radii. Comparison with the value, which I found before, in (3.31), shows that

$$\text{Speed} = \frac{1.095}{2,160} = 5.07 \times 10^{-4} \text{ lunar radii s}^{-1}. \tag{3.40}$$

This is not the same value that I obtained before. Then, in (3.31) I obtained a value of 5.13×10^{-4} lunar radii s^{-1}. I propose to average the two values, to obtain a value

$$\begin{aligned} \text{Speed} &= \frac{5.07 \times 10^{-4} + 5.13 \times 10^{-4}}{2} \\ &= 5.10 \times 10^{-4} \text{ lunar radii s}^{-1} \\ &= 2.55 \times 10^{-4} \text{ lunar diameter s}^{-1}. \end{aligned} \tag{3.41}$$

Please note in passing that the approximate error introduced at this point can be worked out by comparing 5.13×10^{-4} lunar radii s^{-1} and 5.10×10^{-4} lunar radii s^{-1}. The ratio of these two is 1.006, suggesting an error of 0.6%.

How to Create Axes Parallel to Lunar Motion Mathematically

If, like my wife, you do not enjoy playing around with CAD systems, but do enjoy doing calculations, here is a way to find coordinate axis parallel to the direction of motion.

Having fit a line

$$y = ax + b,$$

per (3.32), through the data in Figs. 3.14 or 3.15, using the method above, you note that the gradient of the line in (3.32) is a. It makes an angle $\arctan(a)$ with the x-axis. Let us call the new x-axis x' and the new y-axis y'.

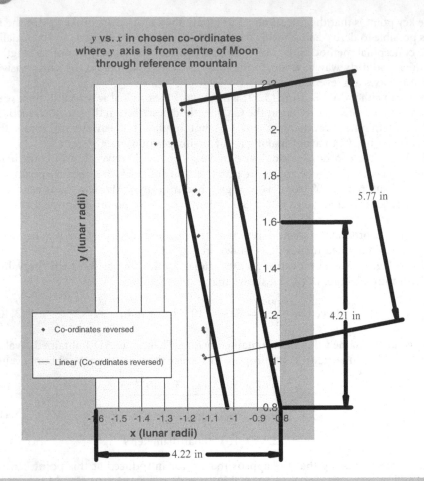

Fig. 3.15. Reversal of coordinates gives the following curve fit. 5.77 in. is the amount of relative movement from the first to the last point, over an interval from $t = 900$ s to $t = 3,060$ s, along an axis parallel to the direction of motion. Since the average of 4.21 and 4.22 in. corresponds to 0.8 lunar radii, my fitting exercise leads me to calculate a relative travel of 1.095 lunar radii.

Then for the ith point

$$x'_i = x_i \cos(\arctan(a)) + y_i \sin(\arctan(a)) \quad \text{and}$$
$$y'_i = -x_i \sin(\arctan(a)) + y_i \cos(\arctan(a)),$$

(3.42)

where the term "arctan" means "angle whose tangent is." "Arctan(a)" is sometimes written "$\tan^{-1}(a)$."

These well-known formulae are derived in elementary texts on calculus such as that by Richard Courant.[20]

Back to the Diameter Calculation: College Mathematics no Longer Required

Way back at (3.12) there was an expression for orbital speed:

$$v_{moon} = \frac{2\pi r_{orbit}}{T_{orbit}} = \frac{2\pi \times 3.810 \times 10^8 \, m}{27.1 \times 24 \times 3,600 \, s} \tag{3.43}$$
$$= 1.0225 \times 10^3 \, m \, s^{-1},$$

where I have used the observed value of 27.1 days for T_{orbit} and (3.16) for r_{orbit}.
Comparison of (3.41) and (3.43) gives

$$2.55 \times 10^{-4} \, \text{diameters s}^{-1} = 1.0225 \times 10^3 \, m \, s^{-1}$$
$$\therefore 2.55 \times 10^{-4} \, \text{diameters} = 1.0225 \times 10^3 \, m$$
$$\therefore 1 \, \text{diameter} = \frac{1.0225 \times 10^3}{2.55 \times 10^{-4} \, m} = 4.010 \times 10^6 \, m \tag{3.44}$$
$$= 4.01 \times 10^3 \, km = 2.47 \times 10^3 \, \text{miles}.$$

How does this compare to published values? The equatorial diameter is 3,476 km or 3.476×10^3 km.

$$\frac{\text{Measured diameter}}{\text{Published diameter}} = \frac{4.01 \times 10^3 \, km}{3.476 \times 10^3 \, km} = 1.154 \approx 1.15. \tag{3.45}$$

Another way to express (3.45) is to say that the measured value is about 15% above the published value.

Well, there is no use denying the discrepancy. Instead, I am going to work my way through the method I used, and winkle out the simplifications and approximations.

Discussion of Quality of Results

Random Errors (Scatter) in Data of Figs. 3.14 and 3.15

The error in the line fitting formula is too tiny to worry about, being about 2.5×10^{-6}%. Interested readers can find out how I calculated this from a statistics book by Box, Hunter, and Hunter.[21]

The Moon's Orbit is Really Elliptical

This actually turns out to be the biggest single source of error. While it is true, (3.10), that the orbital ellipse is very nearly circular, the Earth–Moon center of mass does not orbit around the center, but around one focus as Kepler discovered all those centuries ago. The distances from the focus to the nearest and furthest points in the orbit are

$$d_{\text{pericenter}} = a(1 - \varepsilon) = a(1 - 0.0549) = 0.9451a,$$
$$d_{\text{apocenter}} = a(1 + \varepsilon) = a(1 + 0.0549) = 1.0549a,$$
$$\therefore \frac{d_{\text{apocenter}}}{d_{\text{pericenter}}} = 1.116. \tag{3.46}$$

In (3.46), a is half the major axis. In other words, the distance to the Moon, r_{orbit}, has an uncertainty of $\pm 6\%$, taking 11.6% to be approximately 12%.

The Telescope is not at Earth's Center

The Earth's radius was taken to be r_E = 3,963 km = 3.963×10^6 m. The distance to the Moon, r_{orbit}, was worked out to be 3.810×10^8 m (3.15). The ratio of these two numbers

$$\frac{r_E}{r_{\text{orbit}}} = \frac{3.963 \times 10^6}{3.810 \times 10^8} = 1.04 \times 10^{-2} \approx 1\%. \tag{3.47}$$

This is roughly the error introduced. A full calculation, involving the tilt of the Earth's axis, the plane of the Moon's orbit, the time, etc., would be more complex. The effort would not justify the extra precision. This error is of the order of 1%.

The Earth–Moon Center of Mass is not at the Center of the Earth

Using the usual formula for center of mass

$$\bar{z} = \frac{z_E m_E + z_{\text{moon}} m_{\text{moon}}}{m_E + m_{\text{moon}}}, \tag{3.48}$$

where z is the distance along the line from the Earth to the Moon, and figures for the masses of the Earth and the Moon, I deduce that the center of mass of this system must be roughly 4.6×10^6 m from the Earth's center. (There is a slight technical point here. Where should you choose your zero point along the z-axis? It actually does not matter. Whatever point you choose, you get the same answer. So choose a convenient point such as the earth's center. z_E is then zero.)

As a fraction of r_{orbit}, this is

$$\frac{\bar{z}}{r_{\text{orbit}}} = \frac{4.6 \times 10^6}{3.810 \times 10^8} = 1.2 \times 10^{-2} \approx 1\%. \tag{3.49}$$

Whiteout due to Photographic Overexposure Makes Moon's Surface Uncertain

In particular, the location of the surface is uncertain. Values of the Moon's diameter are compared in two photographs in Fig. 3.16. In one case, the Moon has to be overexposed or Saturn is invisible. The other photograph the Moon is correctly exposed. The apparent lunar diameter is larger when it is overexposed.

The uncertainty is compounded by the fact that in neither the upper nor the lower picture in Fig. 3.16 is the edge of the Moon unambiguously defined. Figure 3.17

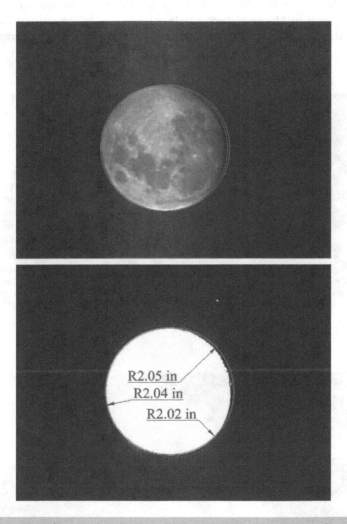

Fig. 3.16. Two photographs taken 1 minute apart. The upper photograph is correctly exposed for the Moon. In the lower photograph, the Moon is overexposed to make the Saturn visible. It can be seen that the apparent diameter of the Moon is larger when it is overexposed.

shows a close-up of the Moon's surface in the nonoverexposed photograph. In this case the surface is diffuse, forcing me to guess where it actually is. The reason for this is probably poor focusing on my part. The method chosen was to use the CAD system to fit circles through three points. The diameters of these circles were then read off.

Figure 3.18 shows a close-up of the Moon's surface in the lower photograph of Fig. 3.16. The surface is much less diffuse in Fig. 3.18 than in Fig. 3.17. The limitation on knowing its position is the size of the pixels.

I estimated the error as follows. The scatter in values of lunar diameter in the nonoverexposed picture in Fig. 3.16 is ±0.01 in 2.01 or ±0.5%. The corresponding figure for the scatter in values of lunar diameter in the overexposed image in Fig. 3.16 is ±0.03 in 2.04 or approximately ±1.5%. Finally, the ratio of medians of overexposed to nonoverexposed diameters is 2.04/2.01 = 1.015. The overestimate is thus 1.5%.

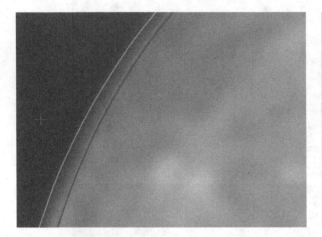

Fig. 3.17.
Closeup from the photograph in Fig. 3.16 in which the Moon is correctly exposed. It can be seen that the Moon's surface is diffuse, forcing us to estimate where it is.

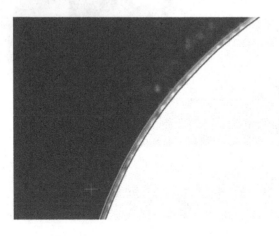

Fig. 3.18.
Closeup from the photograph in Fig. 3.16 in which the Moon is overexposed. The surface is now sharp. The limitation on resolving it is the pixel size.

Variation of Earth's Radius and Gravitational Acceleration

The gravitational acceleration little *g* varies by about 0.5% from Equator to Poles, according to the International Union of Geodesy and Geophysics (http://www.iugg. org). The variation in the earth's radius is about 0.3% according to French. It is partially double-counting to claim these as separate uncertainties, because part of the reason for the weaker gravity at the equator is that the Earth bulges slightly. The remaining difference is due to the centrifugal effect. We therefore expect a 0.5% error.

Combination of Uncertainties

You can usually get one over on your friends by asking them if they know the difference between accuracy and precision. Not many people know. Figure 3.19 shows the difference. Most of the errors discussed above are errors of accuracy. The only exception is the scatter in the two measured lunar diameters. While there are formulae for dealing with the combination of errors of precision, these do not apply to systematic errors of accuracy. Strictly speaking, since the theories underlying the systematic errors such as the noncircularity of the Moon's orbit are well understood, we should be able to calculate the error introduced.

Such an approach is too complex to fall within the scope of this chapter. Instead, I rather cheekily add the percentage estimates of error to give

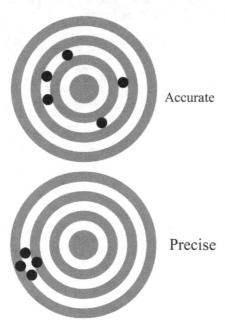

Fig. 3.19. Showing the difference between accuracy and precision. High precision implies high reproducibility. High accuracy implies high "correctness" or conformity to some accepted true value. In real research situations, we often only know the precision.

Accurate

Precise

$$\text{Error}_{\text{Total}} = \text{Error}_{\text{Ellipticity}} + \text{Error}_{\text{center-to-surface}} + \text{Error}_{\text{center-of-mass}}$$
$$+ \text{Error}_{\text{Variable } g} + \text{Error}_{\text{Lunar Diameter}} + \text{Error}_{\text{Line Fitting}}$$
$$= 6\% + 1\% + 1\% + 0.5\% + (0.5\% + 1.5\% + 1.5\%) + 0.6\%$$
$$= 12.6\%.$$

(3.50)

Substitution of this error into (3.44) gives

$$\text{Lunar Diameter} = 4.01 \times 10^3 \,\text{km} \pm 12.6\%$$
$$= (4.1 \pm 0.5) \times 10^3 \,\text{km} \qquad (3.51)$$
$$= (2.5 \pm 0.3) \times 10^3 \,\text{miles}.$$

The range covered by (3.51) lies $15 \pm 12.6\%$ above the published value. There is still a discrepancy. It is therefore not sensible to report the result to a higher number of significant figures.

I am afraid cannot explain the discrepancy away. Furthermore, I am not losing any sleep about it. Too many people tell their story as if everything works smoothly all the time. In my backyard, the telescope goes out of alignment; the computer "freezes up" and stops capturing webcam images just when the sky clears; and I trip over everything in the dark. A certain amount of failure and imperfection is part of the game. If you expect your Solar System projects to work perfectly every time, you will be disappointed, and end up throwing your kit into a dark corner of your garage, only to re-emerge next time you move house. Getting within $15\% \pm 13\%$ is a cause for celebration, not misery. After all, Hubble's great achievement in proving that Andromeda is a separate galaxy was based on a measurement that was over 100% out.[22] My nonastronomer friends are amazed when I tell them I worked out roughly the size of the Moon without leaving home.

CHAPTER FOUR

Jupiter's Moons: Where You Can Watch Gravity Do Its Thing

Why the Jovian System is a Good 'Gravity Laboratory'

Most of the gravitationally driven phenomena in the Solar System are slow. The Moon takes a month to go through its phases; the planets take anywhere from 3 months to 164 years to complete an orbit; and the known Kuiper Belt objects take anywhere from about 250 to over 11,000 years to complete an orbit. You sure cannot see them do much in a night's observing.

Not so the Galilean Moons of Jupiter. With binoculars you can see that they are in very different positions every night. With a telescope and a webcam, you can detect movement of the inner moons in 20 min. Eclipses occur more than once every 2 days, although not necessarily at night from a given position on Earth. Over an evening you can often watch the paths of a pair of moons cross. The one that was nearer to Jupiter ends up being the further.

Some of the Saturn's moons are also fast movers, but they are twice as far away and therefore much harder to see. They are at the limit of what I can photograph with my 8 in. telescope. The Galilean Moons of Jupiter are much easier to observe.

The names of the moons, in order of distance from Jupiter, are Io, Europa, Ganymede, and Callisto.

The Galilean Moons are thus a wonderful "laboratory" in which to see whether Newton's laws of gravity and motion are obeyed. In this chapter, I will show you how you can demonstrate that the Galilean Moons' orbits are very nearly circular and that their orbits are indeed governed by Newtonian mechanics.

J.D. Clark, *Measure Solar System Objects and Their Movements for Yourself!*,
Patrick Moore's Practical Astronomy Series,
DOI: 10.1007/978-0-387-89561-1_4, © Springer Science + Business Media, LLC 2009

I will also show you the amazing fact that the innermost three have orbital periods in the ratio of 1:2:4. To within experimental error, this ratio is exact. In other words, for every orbit Ganymede makes, Europa makes exactly two orbits and Io makes exactly four.

We will barely have scratched the surface of what amateur astronomers could do to study this system of Moons. I will suggest some other measurements that you could make.

My Equipment

I used my 8 in. Newtonian reflector with the webcam of the moment, a Philips SPC900NC^{TM}.

I hesitate to recommend a webcam because it may well be obsolete by tomorrow afternoon, but this particular one has served me well. It is only suitable for the Moon and planets, but I knew that when I bought it. Almost any laptop computer is suitable. I do not use the video capture software that came with the camera. It produces ".mov" files, which at the time of reading cannot be read by the stacking software I use, Registax^{TM}. Translating these to ".avi" files reduces the quality.

Registax can be downloaded for free from http://www.astronomie.be/registax/.

For video capture I use a package called K3CCD Tools^{TM}. This can be downloaded from http://www.pk3.org/Astro. It cost me $50, which I do not regret spending. The output is in ".avi" format, which goes straight into Registax. K3CCD Tools will do the same postprocessing job as Registax, but when I compared them I got better results with Registax. I have learnt a lot about astrophotography since then, so my opinion could well be premature.

My setup is shown in Plate 4.1. I use a flat-packed assemble-it-yourself clear-plastic-coated miniature greenhouse for shelving, because it keeps the dew off my kit, and is fairly mobile. I did learn the hard way to add some guy ropes and tent pegs to combat the wind. Fortunately no lenses broke as it toppled.

Another worthwhile antiwind investment is a stainless steel tripod. It is much stiffer than its aluminum predecessor, not least because that had jointed legs.

Observing in the Rain (Well, Almost)

The month I chose for this project, July 2007, looked like a good idea because I could observe at a fairly civilized time of the night. What I did not bank on was the weather. Most of England had the wettest July since Noah sold his ark for scrap. At one point, large parts of the country ground to a halt because of floods. Nevertheless I got photos on all but two of the 16 nights I needed for the project.

If I had been a little less fastidious about setting up, I would have got photos on one of those two cloudiest nights. The lesson I learned was to have the webcam

Plate 4.1 The author and his telescope. The laptop computer and other accessories are kept on shelving in a flat-pack miniature greenhouse. This keeps the dew and rain off the equipment. The guy ropes keep it stable. You can barely see the trolley under the greenhouse. I move the set-up around my backyard depending on the direction in which I wish to look, and put it in the garage when the weather is bad. The brief case is an old one I use to move the laptop: I never leave that outside! (Picture: Darren Sprunt, used with permission).

recording while I did the setting up, in case it was the only footage I got that night. A few nights later that is exactly what happened.

It is surprising how well the image of Jupiter will penetrate clouds. My 8 in. telescope could detect it well enough to adjust the tracking manually for quite long periods when I could not see it with my naked, or at any rate bespectacled, eye. Indeed the whole exercise was a bit like what I imagine fishing to be: you spend most of your time waiting.

When I was able to take pictures, I would take some shots of Jupiter so that I could use the bands to orient the Moons, and then overexpose the planets in order to see the Moons. With an equatorial mount, in theory the orientation on your computer screen will be preserved all evening until you remove the webcam. With an Alt-Az mount, you would need to take an orientation shot for every picture of the Moons.

This little project led me to a new theory of the Great Red Spot. This is because red spots are what I got covered in from mosquito bites. On all those warm evenings I had to cover my body more thoroughly than in winter. The beginner's astronomy books do not warn you about that. Eventually in a hardware store I found a little gizmo for small fishing boats that scares off the insects so long as there is not too much wind. I can report that inside my little faux-greenhouse it was quite good at keeping insects away from my luminous computer screen, which did rather attract them. A traditional Insect-o-Cutor™ type bug zapper is totally unsuitable for astronomy because the bluish light destroys whatever dark-adapted vision is left after your laptop has dazzled you.

K3CCD Tools turns your PC screen from color to varied hues of dark red. Unfortunately Registax still shows large areas of white even trough this. I am a great believer in processing the image as soon as possible in case you need to take another, but the software does not do your dark adjustment any good. When I politely complained to the Registax team about this, within a couple of hours I got a very helpful reply about their plans for the next version, and asking me to show them in detail what I mean. You cannot ask for more than that.

A typical photo I took looks like Fig. 4.1. With practice it is easy to recognize the outer two Moons, Ganymede and Callisto. Ganymede is much the brightest; and

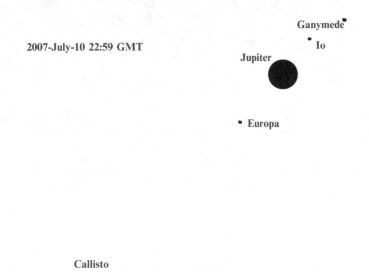

Fig. 4.1. Jupiter's Moons, taken from a typical photo. To make the picture cheaper to print, I turned the photo into a negative and turned the contrast right up so that there are no shades of gray. The apparent size of the Moons is an illusion. They are much smaller than they appear. What you see is the effect of limitations in both optics and webcam. I also tilted the webcam to put the Moons across the diagonal. Otherwise I would have had to make a montage of two photos. When the outermost Moon, Callisto, was at its widest apparent separation from Jupiter, I had to do this anyway.

Callisto is the least bright of the four. By playing with the exposure level on your webcam, you can soon figure out which these are. Most of the time, they are further away from Jupiter, which helps.

The inner two, Io and Europa, are much harder to tell apart. Io has the higher angular velocity as it orbits, so it moves faster than Europa, enabling you to guess which is which on sequences of photographs; but it is not always easy. You can always cheat and look in the astronomy magazines. Many of them publish monthly charts of the positions of Jupiter's Moons' movements when they are visible at night. Sky and Telescope^{TM} even publish an "applet" on their Web site, which can be found by clicking on the "observing celestial objects" link on http://www.skyandtelescope. com. As usual with Web sites, it could get moved in a future update of the site, so you may need to hunt for it.

Measuring Angular Separations on Photos

Once again, I used computer-aided drafting (CAD) software to make measurements from the photographs. The result for the photo in Fig. 4.1 is shown in Fig. 4.2.

The measurements are in inches. This is an arbitrary measurement: I could choose what size I pasted the photograph into the CAD file. What I really want, though, is to convert the size into arcminutes, i.e., the angular separation.

To provide a rigorous calculation of the angular separation in all cases from the simplest convex objective lens to a Schmidt-Cassegrain telescope with its complex corrector and twin mirrors would be the subject of a book in its own right. For a simple convex objective lens, the calculation is relatively simple, and runs as follows.

An objective lens works by collecting rays of light from an object at the point such as A in Fig. 4.3 and focusing them to the point B in Fig. 4.3. Of course, the planet and its satellites shown are much, much further from the lens than the focal point at B. Indeed, they are about a trillion times as far, but showing that would make a properly scaled diagram slightly too big for the page. All the rays going from A to B are bent (i.e., refracted) except one. The ray going through the center of the lens travels in a straight line.

It is obviously easier to follow the geometric behavior of straight lines than ones bent by refraction. Therefore, I am going to analyze the rays which pass through the center of the lens.

In Fig. 4.4 the object is assumed to be in the middle of the view field. You can always point the lens so that this is true, so it is not a restrictive assumption. The tangent of $\theta/2$ is given by

$$\tan\left(\frac{\theta}{2}\right) = \frac{(w/2)}{f} = \frac{w}{2f}, \tag{4.1}$$

since tangent = opposite/adjacent. In practice θ is likely to be a very small angle. The largest-looking Solar System objects in the sky, the Sun and Moon, are only about

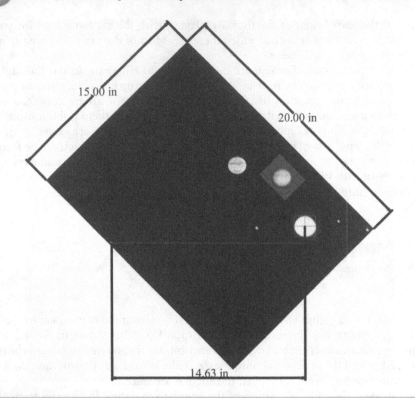

Fig. 4.2. How I used CAD software to make measurements. First, I superimposed my orientation shot. In this case, I only managed one picture all night, but had two nonoverexposed orientation photos. I "eyeballed" one of them for orientation. This procedure proved to be accurate to 1°. I then used the rectangular *box* to line up the edges of the overexposed Jupiter. The CAD software provided the means to place the *lines* through the midpoints of the *rectangle*. These *lines* then passed through the center of the over-exposed *white blob* representing Jupiter. I drew *horizontal lines* from this mid-line to each of the Moons, making full use of the zoom capability of the CAD software to place these lines as close to the centers of the Moons as I could estimate. For clarity, only the line to Callisto is shown. The CAD software measured all the distances in inches. I measured the size of the whole photo, because I knew the size of the photo in arcminutes. This gave me my conversion factor from inches to arcminutes.

half a degree across. If we measure θ in radians, an approximation (see Appendix A) can be used:

$$\tan(x) \approx x \quad \text{if } x \leq 0.1 \text{ radians.} \tag{4.2}$$

For $x = 0.1$ radians, the error in (4.2) is no more than three parts in a thousand. Substituting (4.2) into (4.1) gives

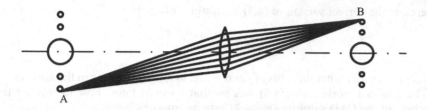

Fig. 4.3. An objective lens works by collecting rays of light from an object at the point A and focusing them to the point B. Of course, the planet and its satellites shown are much, much further from the lens than the focal point at B. All the rays going from A to B are bent (i.e., refracted) except one. The ray going through the center of the lens travels in a straight line.

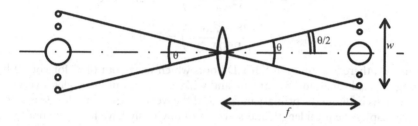

Fig. 4.4. In this diagram, rays from the extremities of the object being viewed are followed. The rays shown are the ones that go through the center of the lens. The size of the focused image of the object is w. The focal length of the lens is f. The angle subtended by the viewed object is θ.

$$\tan\left(\frac{\theta}{2}\right) \approx \frac{\theta}{2} = \frac{w}{2f};$$
$$\therefore \theta = \frac{w}{f}; \tag{4.3}$$
$$\therefore w = f\theta.$$

In other words, if we know two of w, f or θ, we can work out the third from (4.3). It is more common to express the angular separations of Jupiter's Moons in minutes of arc than in radians. Remembering that there are 60 min (or 60′) in 1°; and 360° or 2π radians in a complete circle, we can work out that

$$2\pi \text{ radians} = 360° \times 60' = 21,600'.$$

$$\therefore 1' = \frac{2\pi}{21,600} \text{ radians} = 2.909 \times 10^{-4} \text{ radians}; \quad \text{and} \tag{4.4}$$

$$1 \text{ radian} = \left(\frac{21,600}{2\pi}\right)' = 3,438'.$$

Therefore the correct version of (4.3) in minutes of arc is

$$\theta = \left(\frac{3{,}438w}{f}\right)'.\tag{4.5}$$

It does not matter what the units of f and w are, as long as they are in the same units.

The reason I worked out (4.5) was so that I would know how to convert the "inches" of my CAD software (Fig. 4.2) into arcminutes.

In Fig. 4.2, the image size was 20×15 in. From the data sheet for the Sony ICX098BQ chip in my Philips SPC900NC webcam,[23] I can get values of w in mm. The chip has 640×480 pixels, each 0.0056 mm apart, according to the data sheet. Multiplying up, the sensor chip has a size of 3.584 mm $\times$ 2.688 mm. The focal length f of my telescope's primary mirror is 1,200 mm. Substituting into (4.5) gives me an area of

$$\begin{aligned}
\theta_{width} &= \left(\frac{3{,}438 \times 3.584}{1{,}200}\right)' = 10.27' \quad \text{and} \\
\theta_{height} &= \left(\frac{3{,}438 \times 2.688}{1{,}200}\right)' = 7.70'.
\end{aligned}\tag{4.6}$$

Fortunately, there is a utility in K3CCD Tools which works out (4.6) for you. It has preprogrammed values for my camera, and will even work out the angles if you have a Barlow lens between the primary mirror and the webcam. A $4\times$ Barlow lens acts as if it quadruples the focal length and so on. You obviously have to tell it about your telescope. You have to tell K3CCD Tools about your webcam in any event to get it to capture the images.

Anyway, my CAD system measured a field of view of $10.27' \times 7.70'$ as 20×15 in. That works out as 0.513 in. per minute of arc

$$1' = \frac{10.27'}{20} = \frac{7.70'}{15} = 0.513 \text{ in. arcmin}^{-1}.\tag{4.7}$$

In theory, (4.7) has to be worked out for each image. In practice, when you paste the photographs into the CAD system, you can usually use the "grid" features in your CAD system to make sure the photos are always pasted in the same size.

Equation (4.7) can be used to convert the distances of the Moons from Jupiter into arcminutes. For example, in Fig. 4.2, the 14.63 in. from Jupiter to Callisto corresponds to an angular separation of $14.63/0.513 = 28.50'$.

Why does the above optical analysis not work for any but the simplest objective lenses?

Achromatic and apochromatic refractors have compound lenses. Any off axis light rays that pass through the center of the first part of the lens will not quite pass through the center of the second and subsequent elements of the lens. Therefore they will experience at least some refraction. The analysis is thus only approximately correct. That does not make it invalid. All scientific theories and analyses are inexact to some extent. There is a fine art to choosing useful approximations to gain insight.

Rays of light never reach the center of the primary mirror in any kind of reflecting or catadioptric telescope, because the center is obscured by a secondary mirror. Hence the analogous analysis for a mirror needs to be done more carefully. It is easy to convince yourself that (4.5) must be right by imagining that by some miracle the secondary mirror lets light into the telescope unmolested, but reflects light from the primary mirror. Easier yet, you could imagine that there is no secondary mirror, and that instead the image is captured on a perfectly transparent chip in a perfectly transparent webcam on the primary optical axis.

Figure 4.5 shows how a primary mirror works by collecting rays of light from an object at the point A and focusing them to the point B. Of course, the planet and its satellites shown are much, much further from the mirror than the focal point at B. All the rays going from A to B have equal incident and reflected angles. If the mirror is aberration free, the rays will converge precisely onto the point B. Otherwise, they will approximately converge on Point B. Now consider (Fig. 4.6) the ray that passes from A to B in Fig. 4.5 via the center of the mirror. Its geometry is the easiest to analyze. Its incident and reflected angles are $\theta/2$. Trigonometric analysis gives $\tan(\theta/2) = (1/2)w/f$ as in the case of an objective lens.

How do we generalize the argument to a real Newtonian telescope? Even if the ray from A to B is interrupted by a secondary mirror, all the other rays shown in Fig. 4.5 approximately converge on the point B. So the image B will be at the same place whether or not there is a secondary mirror which blocks the rays which would have landed closest to the center of the primary mirror. Therefore the value of w is unaffected by the presence of a blocking secondary mirror.

If the rays are reflected sideways by the secondary mirror, as is the case with Newtonian and Dobsonian telescopes, the length f from the center of the primary mirror to the focal plane is not changed. Nor is the value of w. Hence (4.1) is also valid for Newtonian telescopes.

Catadioptric telescopes are unfortunately more complex to analyze.

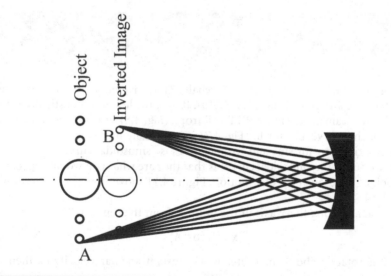

Fig. 4.5. A primary mirror works by collecting rays of light from an object at the point A and focusing them to the point B. Of course, the planet and its satellites shown are much, much further from the mirror than the focal point at B. All the rays going from A to B have equal incident and reflected angles.

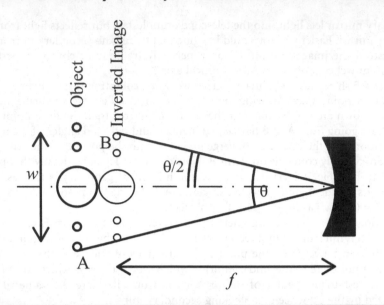

Fig. 4.6. Now consider the ray that passes from A to B in Fig. 4.5 via the center of the mirror. Its geometry is the easiest to analyze. Its incident and reflected angles are $\theta/2$. Trigonometric analysis gives $\tan(\theta/2) = (1/2)w/f$ as in the case of an objective lens.

Making Sense of the Moon's Positions

Looking at these data, it is easy to "eyeball" and imagine a sine or cosine curve through the positions of Callisto (Fig. 4.10). It is a little less easy, but still not difficult to do this for Ganymede (Fig. 4.9). For Europa (Fig. 4.8), it is getting quite hard to imagine such a curve, but for Io (Fig. 4.7) it is virtually impossible.

The generic name for sine and cosine curves is "sinusoidal curves."

The reason for the sinusoidal curves is that they are what you would expect if you were looking edge-on at circular motion. Figure 4.11 shows how circular motion can be resolved.

The distance along the x-axis of the point P from the center is

$$x = r\cos(A). \tag{4.8}$$

But if P is rotating about the center with constant angular velocity ω, then (4.8) becomes

$$x = r\cos(\omega(t - t_0)) = r\cos(\omega t - \varphi), \tag{4.9}$$

where t_0 is a time in the past at which the angle A was zero, and the Greek lower case letter phi (φ) has the meaning

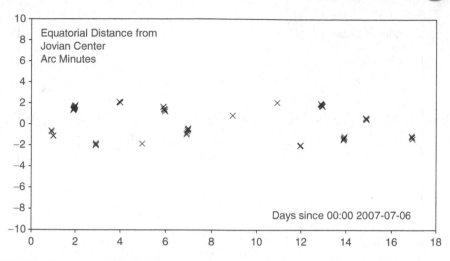

Fig. 4.7. Data collected from 6 to 20 July 2007 of measurements of the positions of Io, measured in arcminutes from the center of Jupiter along lines parallel to Jupiter's equator.

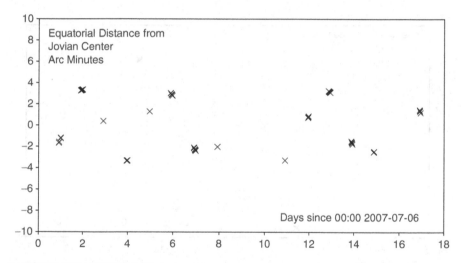

Fig. 4.8. Data collected from 6 to 20 July 2007 of measurements of the positions of Europa, measured in arcminutes from the center of Jupiter along lines parallel to Jupiter's equator.

$$\varphi = \omega t_0. \tag{4.10}$$

My objective will be to find a curve like (4.9) for each moon, which fits the orbital data shown in Figs. 4.7–4.10. Once I have done this, I will know the values of r, ω, and φ for each of the Galilean satellites. That is enough to specify a circular orbit completely (Fig. 4.11).

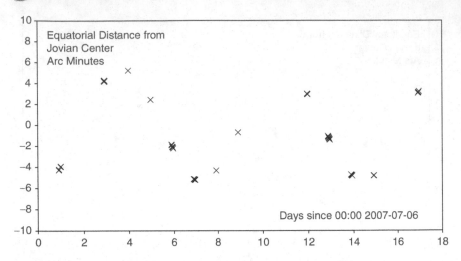

Fig. 4.9. Data collected from 6 to 20 July 2007 of measurements of the positions of Ganymede, measured in arcminutes from the center of Jupiter along lines parallel to Jupiter's equator.

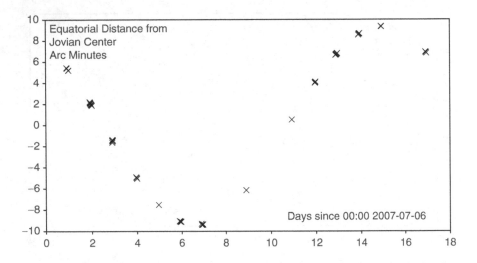

Fig. 4.10. Data collected from 6 to 20 July 2007 of measurements of the positions of Callisto, measured in arcminutes from the center of Jupiter along lines parallel to Jupiter's equator.

If the orbits deviate significantly from circularity, the method will not work, and I will not get good fits to (4.9).

So let us see how well I do.

I could not find a method in any of my textbooks or on the Internet for fitting sine curves, so I had a go at inventing one for myself. My wife was unimpressed by my

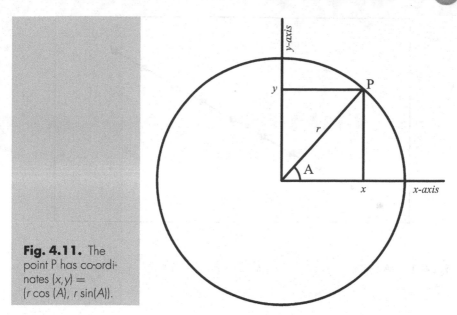

Fig. 4.11. The point P has co-ordinates $(x, y) = (r \cos (A), r \sin(A))$.

method. At first she called it "cack-handed." In Britain, that is a derogatory term for left-handers. She being American did not know that, but I, being both English and left-handed, certainly did, and was suitably miffed. My dander was up. Somebody needed to be proved wrong, and I hoped it was not going to be me.

As mentioned in [Chap. 3], the traditional way to fit straight lines, i.e., mathematical functions $y(t)$ of the form

$$y(t) = at + b, \tag{4.11}$$

where a and b are constants, through data is the "least squares method." $y(t)$ means that the value of y depends on the value of t. We may think of t as "time" for present purposes. Then $y(t)$ means that the value of y depends on the time.

Figure 4.12 shows how a variable such as $y(t)$ increases uniformly with time. Figure 4.13 shows what might happen if I do not know $y(t)$ perfectly, due to the limitations of my measurement technique. In contradistinction, I know the time at which the measurement was taken very accurately. For example, K3CCD tools logs the time at which webcam sequences are recorded; and the laptop regularly updates its time over the Internet whenever it is connected.

Let me call the uncertainty in the ith measurement of y, taken at time t_i, Δy_i.

Methods that minimize the sum of the squares of Δy_i, (see Appendix B) so that

$$\sum \left(\Delta y_i^2 \right) = \min, \tag{4.12}$$

are called "least squares" methods. The idea is to calculate or estimate the straight line in Fig. 4.13 that satisfies (4.12).

The formulae for doing this for the case where $y(t)$ increases, or is thought to increase, linearly with time, are widely published. You can find them in many elementary statistics textbooks, such as the one by Spiegel, Schiller, and others.[24]

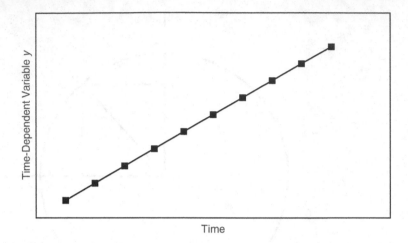

Fig. 4.12. A straight line of the form $y(t) = at + b$, like 4.11. $y(t)$ increases uniformly with time.

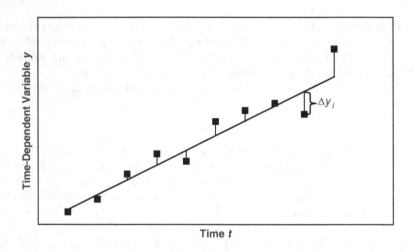

Fig. 4.13. A straight line of the form $y(t) = at + b$, like (4.11). $y(t)$ increases uniformly with time. But there are now apparently "random errors" in the values of $y(t)$ due to imperfections in the measurement technique.

As an undergraduate, I was troubled by this choice of least squares as a criterion for "fitting" or averaging a line. It was apparently first suggested by the great mathematician C. F. Gauss (1777-1855),[25] who seems to have dreamt it up as a teenager. Gauss proved that the least squares procedure is tantamount to assuming that the errors Δy_i are distributed like a "normal" of "Gaussian" distribution (Fig. 4.14); and I show this in Appendix B.

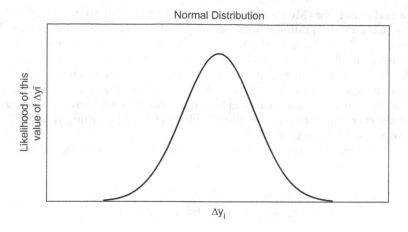

Figure 4.14. A normal distribution of values of Δy_i, which are said to be "normally" distributed.

This is emphatically an assumption, not a fact. I show in Appendix B that normally distributed errors are the most likely state of affairs, assuming that there are no systematic errors present in the data. This is what Gauss showed.

What I wanted was an analogous method to calculate a least-squares cosine, per (4.9), rather than a least squares line.

I drew a complete blank trying to adapt the procedure given by Spiegel, Schiller, and colleagues[24] to sinusoidal curves. Maybe I am not smart enough. Instead, I invented a procedure to guess the answer. It is illustrated in Appendix B. Well, I did not completely invent the method – it is one of a class of methods known to mathematicians as "Monte Carlo" methods.[26] Monte Carlo methods take advantage of computer-generated random numbers to simulate a phenomenon subject to variation. What you do is to rerun several thousand instances of the simulation, and see if you tend to get the same answer. If you do, you conclude that the simulation has done a good job of estimating the answer.

There are weaknesses of Monte Carlo methods. First, there is no guarantee that they get the right answer. They represent the triumph of optimism over certainty. Second, you have to run thousands upon thousands of simulations before the output "converges" to a fairly constant answer. In other words, they are inefficient methods. However, it is a computer doing the drudgery, not me, so what do I care if it takes a while? All I care about is that the time the computer takes is not hard on my patience. Modern computers are not exactly slow.

What I did was to take (4.9) and modify it as follows.

$$x = (r_{\text{trial}} + r_{\text{random}})\cos((\omega_{\text{trial}} + \omega_{\text{random}})t - (\varphi_{\text{trial}} + \varphi_{\text{random}})). \qquad (4.13)$$

In other words, I modified each of the three unknown variables r, ω, and φ by starting with a trial value and adding a random number to it. The random number could have been positive or negative.

I actually did the Monte Carlo runs of (4.13) in Microsoft Excel™. This package has a random number function *rand( )* which generates a random number between 0 and 1. I have in the past run checks to see how evenly distributed the random numbers are. I would check the distribution of 30,000 or so values of *rand( )*, and could find no evidence of a systematic bias. To use this function in Excel, you simply type "= rand( )" into the cell. You will find that every time you do a calculation anywhere else in that Excel file, your value of *rand( )* resets itself. You can also make every instance of *rand( )* in the file reset itself by hitting your "F9" key. This takes some getting used to.

First, however, let me show you how I wrote formulae to randomize r, ω, and φ. In the case of the orbital radius r, I would write formulae like

$$(r_{trial} + r_{random}) = r_{trial}(0.95 + 0.2\,\text{rand}( )) \quad \text{or}$$
$$(r_{trial} + r_{random}) = r_{trial}(0.995 + 0.02\,\text{rand}( )). \tag{4.14}$$

In the top line of (4.14) I have allowed r to vary by $\pm 5\%$ about its trial value; and in the second lone I have allowed it to vary by $\pm 0.5\%$ about its trial value. I did not know how much it would have to vary, so I played around to see what would happen.

Similarly, the variables ω and φ were varied like this:

$$(\omega_{trial} + \omega_{random}) = \omega_{trial}(0.95 + 0.2\,\text{rand}( )) \quad \text{or}$$
$$(\omega_{trial} + \omega_{random}) = \omega_{trial}(0.995 + 0.02\,\text{rand}( )). \tag{4.15}$$

$$(\varphi_{trial} + \varphi_{random}) = \varphi_{trial}(0.95 + 0.2\,\text{rand}( )) \quad \text{or}$$
$$(\varphi_{trial} + \varphi_{random}) = \varphi_{trial}(0.995 + 0.02\,\text{rand}( )). \tag{4.16}$$

From (4.13), the squares of the differences at any given time are

$$(x_{observed} - x_{fitted})^2 = (x_{observed} - (r_{trial} + r_{random})\cos((\omega_{trial} + \omega_{random})t$$
$$- (\varphi_{trial} + \varphi_{random})))^2. \tag{4.17}$$

What we are aiming to do is to find the fits that minimize the sum of these squares of differences

$$\sum (x_{observed} - x_{fitted})^2 = \sum (x_{observed} - (r_{trial} + r_{random})\cos((\omega_{trial} + \omega_{random})t$$
$$- (\varphi_{trial} + \varphi_{random})))^2 = \text{lowest possible value}. \tag{4.18}$$

What you do is to calculate the difference equation (4.17) at every point you have observed, square it and add them all up, as in (4.18). You do this several thousand times, store the results; and note which one gives the lowest sum of squares of differences.

If you do this in Microsoft Excel, you will find that as you search for the run that gives the minimum, the *rand( )* function keeps recalculating. You can then never find what you want. There is a simple workaround. Select the whole worksheet, copy it and paste it into cell A1 of a blank worksheet. But do not use normal paste. Instead use "Paste

Special" and select "Values." You can then immediately do a second paste special and select "Formats," and the formatting from the copied sheet will be transferred. The new sheet will contain only numbers, not functions, so it will never recalculate itself. You can then search it for the smallest sum of squares of differences.

How many times do you have to run the Monte Carlo simulation? A lot. I kept doubling the number until I got reproducible best fits. Previous experience with other Monte Carlo simulations told me that I was going to need somewhere around $2^{14} = 16,384$ repeats. I actually settled on $2^{15} = 32,768$ repeats.

Because of the recalculating problem in Excel, the way I ran my 32,768 repeats was to run $2^{13} = 8,192$ runs in one worksheet, copy and "paste-special" into another; and repeat four times, so that my results were "paste-specialled" as values into four worksheets. This cut down the waiting time while the wretched *rand( )* kept recalculating itself in several cells.

There was no one "correct" or "best" percentage by which to vary all of these variables. I played it by ear. I started with trial values, and used those to repeat the process, each time reducing the percentages of r, ω, and φ which were allowed to vary randomly. In the case of Callisto, I found that I had good fits to ω and φ, but a poor fit to r. So I held ω and φ constant and reran the Monte Carlo simulation with only r randomized.

How did I choose my starting values of r, ω, and φ? I set up a spreadsheet where I could change the starting values and look at what happened to the sinusoidal curve with my data showing. By the time I had collected the data, I had already seen for myself that the published orbital periods were at the very least about right. After all I had been watching these moons closely for 20 days; and had been observing them haphazardly for several months before that. So I used the published orbital periods from Wikipedia,[27] and the formula $\omega = T/2\pi$ to relate orbital period to angular velocity.

The values of r are not often published in arcminutes. However, I did use the Sky and Telescope Internet applet referred to earlier to work out when the moons would be at maximum amplitude. I got lucky with the weather and managed to get photographs not actually at peak amplitude, but close enough for me to measure starting values of r for each moon (Figs. 4.15 and 4.16).

The starting values of φ for each moon I guessed by fiddling around in Microsoft Excel until I could "eyeball" good values by looking at plots made with various assumed values.

The answers I got seemed to be reproducible: it did not matter much what I chose as starting values for r, ω, and φ. The method would find very similar final values. As with all trial-and-error mathematical methods, you keep going until you are happy with your result. This is always a subjective decision. Do not let anyone ever bamboozle you into thinking otherwise!

Figure 4.17 is evidence that my curve fitting method works; and that my wife's insulting remark about it should be diplomatically forgotten in the interest of marital harmony. Successful husbands do not do gloating. Nor do I.

This same figure can also be used to look for noncircularity in the orbits. If there were significant ellipticity, we would expect the sine curves to be poor fits, and to see systematic evidence of faster travel in the orbit when the moon is closer to Jupiter, and slower travel when it is further away. I cannot see any such bias in Fig. 4.17. From this I conclude that the accuracy of the approximation of circular orbits is at

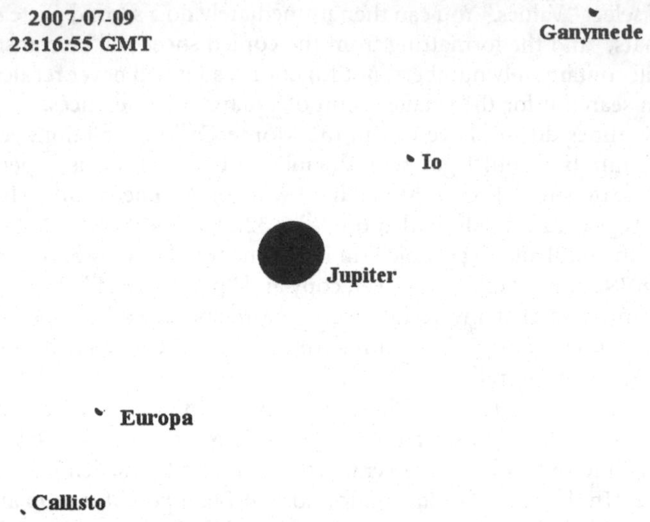

2007-07-09
23:16:55 GMT

Ganymede

Io

Jupiter

Europa

Callisto

Fig. 4.15. In this photo, shown as a negative, Io, Ganymede, and Europa are all at almost their maximum angular separations from Jupiter. This gives us trial values for *r* for each of these Moons. Incidentally, the little "tufts" sticking out of the moons are a sign that my telescope was out of collimation. A key principle of experimental science is that you should always report what you actually observed, not what you should have observed, however, embarrassing this may be.

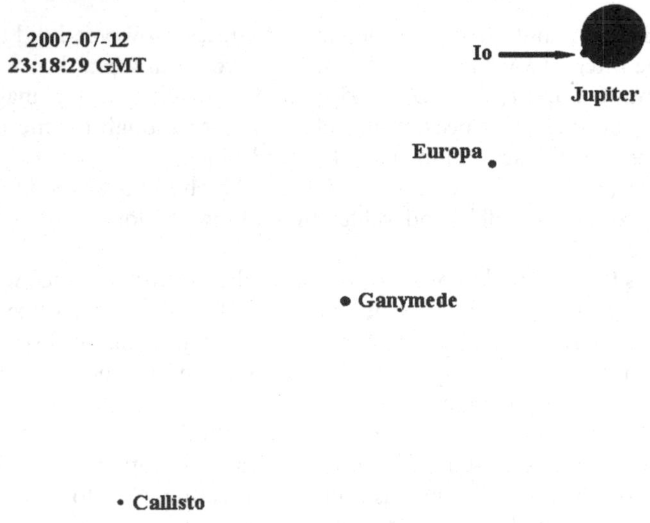

2007-07-12
23:18:29 GMT

Io

Jupiter

Europa

Ganymede

Callisto

Fig. 4.16. In this photo, shown as a negative, Callisto is at almost its maximum angular separations from Jupiter. This gives us trial values for *r* for each of these Moons. Incidentally, Io is beginning to cross in front of Jupiter. A few nights earlier, I had witnessed the eclipse end as Io came out from behind Jupiter's shadow on that side of the planet. Sadly I did not capture this event as webcam footage. But it is the reason why I know that Io is in front of Jupiter at this time, not behind it.

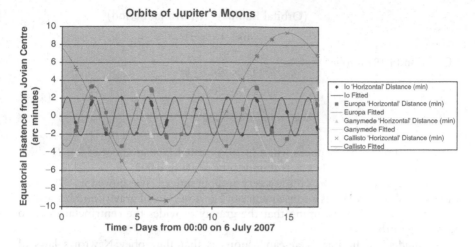

Fig. 4.17. The fitted curves from (4.13) compared against the measured positions of the Galilean Moons of Jupiter.

Table 4.1 Values of r, ω, and φ Used to Plot Fig. 4.17

Moon	Fitted Amplitude i.e., Orbital Radius (arcmin)	Fitted Angular Velocity (radians/ day)	Predicted Orbital Period (days)	Published Orbital Period[27] (days)	Fitted Phase Angle φ (radians)
Io	2.139	3.544	1.77	1.77	−7.557
Europa	3.354	1.770	3.55	3.55	−3.667
Ganymede	5.399	0.878	7.16	7.16	−3.237
Callisto	9.379	0.382	16.47	16.69	0.609

least as good as the accuracy of my measurements. In other words, if the orbits are elliptical, my measurements did not detect this.

Close examination of Fig. 4.17 and Table 4.1 shows that the orbits of Io, Europa, and Ganymede appear to be related, but that the orbit of Callisto is not obviously "tracking" those of the other three.

The orbits of Io, Europa, and Ganymede have periods in the ratio 1:2:4. Their phases are also related: when Ganymede is at maximum separation from Jupiter, so are the other two. Furthermore, they are never all on the same side of Jupiter when their apparent angular separations from it maximize. I have observed for myself that all the above is at the very least approximately true.

I am of course not even almost the first person to notice that these three moons' orbits track one another. This phenomenon is known as a "Laplace resonance."[28]

Another question in which I was much interested was: could I use my data to test Kepler's third law of planetary motion, that

$$\text{(Orbital Radius)}^3 \propto \text{(Orbital Period)}^2$$

$$\text{or} \quad r^3 \propto T^2. \tag{4.19}$$

Equation (4.19) implies that

$$\log r^3 \propto \log T^2 \text{ or}$$

$$3\log r \propto 2\log T \tag{4.20}$$

$$\text{i.e.,} \quad \log T \propto \frac{3}{2}\log r.$$

Figure 4.18 shows what happens when (4.20) is tested. The gradient of $\log T$, which (4.20) predicts to be exactly 1.5, is found to be 1.504 ± 0.002.

The discrepancy is 0.27%. This is a very good fit indeed. We showed in [Chap. 1] that Kepler's third law, (4.19), is a consequence of Newton's law of gravity and his second law of motion; and of assuming that the gravity provides the centripetal force to maintain an orbit.

My "model" of Jupiter's Galilean Moons is that they obey Newton's laws of motion and of gravity, and that they orbit in circles. A necessary condition that all

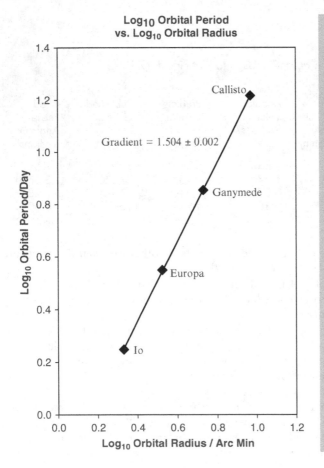

Fig. 4.18. A plot of the logarithm of orbital period vs. the logarithm of orbital radius.

this be (approximately) true is that Kepler's third law, (4.19), be valid when my values of r are used. To within 0.27%, it is.

This does not prove that my model is right; but it does strongly suggest that it is a good model. That is about as good as scientific proof ever gets.

Why do the moons of Jupiter not lie on a straight line as we look at them?

This question had been puzzling me for some months before I undertook my project to track these moons. I did not particularly expect to discover the answer, but I did. There is in fact a very simple explanation.

If we were looking at the orbital plane from the poles of Jupiter, the orbits would describe the paths shown in Fig. 4.19.

If in July 2007, we on Earth were looking exactly toward the plane of the equator, the moons would all be collinear with Jupiter's equator. But we were not. Figure 4.20 shows the orbits approximately as I observed them at that time. I mainly estimated this from the positions of the outer moon Callisto on my photos over the time of the project.

This explains why the moons of Jupiter do not appear to be collinear. We are looking at their orbital plane almost but not quite edge-on.

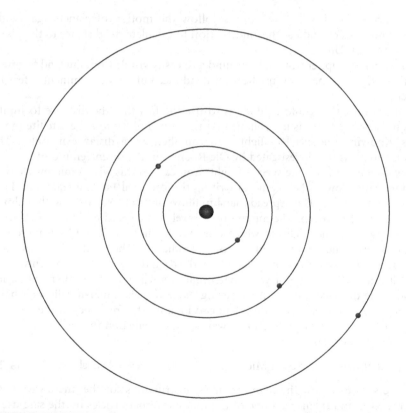

Fig. 4.19. The orbits of Jupiter's Moons as seen from one of Jupiter's poles. The orbital radii are to scale, as is Jupiter itself, but the sizes of the moons have been exaggerated to make them visible.

Fig. 4.20. The orbits of Jupiter's Moons approximately as seen from my back yard in July 2007. The orbital radii are to scale, as is Jupiter itself, but the sizes of the moons have been exaggerated to make them visible.

Strictly speaking I have assumed that they all orbit in the same plane.

For a start, you could check my assumption that the moons all orbit in the plane of Jupiter's equator.

Other Jupiter Projects You Could Try

In Chap. 5 and 6, I will show how to follow the motion of planets against the background of stars, and use this information to calculate the distance to the planet in Astronomical Units.

Unfortunately Jupiter will not be around and easily visible from England for me to do this in time to meet my publication deadline, but it is an eminently feasible project.

Another thing that could be done is to use the fact that the distance to Jupiter varies because the Earth is in orbit around the Sun; and to time the satellite movements. Knowing the speed of light, you can then get a distance in miles. This phenomenon was first investigated by Ole Roemer in the seventeenth century.[29]

How do you estimate the speed of light? You can do this with a microwave oven full of marshmallows.[30] No, I am not writing this on April 1st. You really can. First, remove the turntable. Then spread marshmallows around a square dish that almost fills the floor of the oven. The microwaves travel at the speed of light. (They are in fact just low-frequency light, to which our eyes are not sensitive.) In a microwave oven, you get "standing waves" of microwave radiation. The amplitude at any point varies from zero to a maximum. That is why the darn things cook so unevenly. What you will find is that some of your marshmallows will melt first, after only a few seconds. The distance between neighboring pairs of melted marshmallows is half a wavelength of microwave radiation. You can look up the frequency of the "magnetron," the device that makes the microwaves, on the label on the back of the oven. Then you use the formula

$$\text{Speed of Light} = \text{Speed of Microwaves} = \text{Frequency} \times \text{Wavelength}. \quad (4.21)$$

Please be sure to follow the manufacturer's instructions for the microwave oven. If you cannot find them, read one of the many Internet articles on the safe use of microwave ovens.

We could then use (1.16) of Chap. 1 to obtain the mass of Jupiter. You would need to know G. The well-known way to measure the mass G is the experiment of

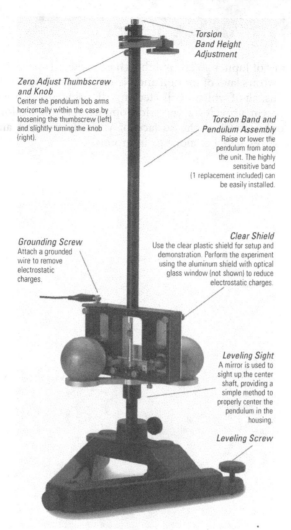

Torsion
Band Height
Adjustment

Zero Adjust Thumbscrew
and Knob
Center the pendulum bob arms
horizontally within the case by
loosening the thumbscrew (left)
and slightly turning the knob
(right).

Torsion Band and
Pendulum Assembly
Raise or lower the
pendulum from atop
the unit. The highly
sensitive band
(1 replacement included) can
be easily installed.

Grounding Screw
Attach a grounded
wire to remove
electrostatic
charges.

Clear Shield
Use the clear plastic shield for setup and
demonstration. Perform the experiment
using the aluminum shield with optical
glass window (not shown) to reduce
electrostatic charges.

Leveling Sight
A mirror is used to
sight up the center
shaft, providing a
simple method to
properly center the
pendulum in the
housing.

Leveling Screw

Fig. 4.21. The Pasco™ version of Cavendish's experiment to measure the universal gravitational constant G. The principle is that the small balls are held by an extremely thin wire, and can swing toward the large balls because of the tiny mutual gravitational interaction.

Cavendish, in which a very fine torsional balance is used to measure the movement of balls of a dense metal a few inches in diameter due to each others' gravitational attraction. A version of this experiment can now be bought at a price which, while beyond most individuals, would be within the scope of an established astronomy club.[31] The apparatus is shown in Fig. 4.21.

Conclusion

The Galilean Moons of Jupiter go around their host planet in approximately circular orbits, obeying Newton's laws of motion and Newton's law of gravity. I have laid out the evidence for this, all of which I collected myself, using amateur equipment.

This satellite system provides a very rich opportunity for exploration. In this chapter, I have only scratched the surface of what determined amateurs could discover for themselves, whether alone or in groups.

CHAPTER FIVE

Sunrise, Sunset

It is remarkable what you can learn from sunrises and sunsets if you keep your wits about you.

First, I will show you how I measured the length of a day on Mars. My webcam, 8-in. *f*/6 telescope, and a 4× Barlow lens that turned out to be 3.4×, were quite good enough to tackle this problem close to the 2007 opposition.

Then I will take you in search of the Sun. Except at total solar eclipses, once the Sun rises, it very quickly dawns on you that you cannot see where the Sun is relative to the stars, so you have to use indirect methods. I had not planned to use the method I did. In fact, I was a bit slow to realize that you need to know where the Sun is to work out the orbits of superior planets.

You have to track the motion of the superior planets relative to the stars for a period of time, and do some calculations. You sure as heck cannot do this without making some decisions about how you describe the positions of the stars. The middle part of this chapter gives an account of the coordinate systems I used. If you have forgotten – or never knew – about Cartesian and spherical polar coordinates, there is a guide in the Appendix A.

The second part shows how you can work out where the ecliptic is. The definition of the ecliptic makes this look easy. The ecliptic is the imaginary line along which the Sun appears to move as the Earth orbits it.

I read somewhere that one of the tricks the early astronomers used to solve this problem was to see what was diametrically opposite the sun at sunrise and sunset, and use the position of that object to back-calculate the track of the Sun. Not in my backyard, you cannot. I cannot see the horizon in my suburban environment. Great! Now what!?

Pause three weeks while I think what to do about this. Inspiration eventually struck; and I will show you how I found out where the ecliptic is.

But first let us have a look at Mars.

J.D. Clark, *Measure Solar System Objects and Their Movements for Yourself!*,
Patrick Moore's Practical Astronomy Series,
DOI: 10.1007/978-0-387-89561-1_5, © Springer Science + Business Media, LLC 2009

A Mars Day for Work, Rest, and Play

Mars rotates. This is one of the first things you notice when observing it. Although the Moon also rotates, it always points much the same face at us. In contradistinction, Mars does not always show the same face to us. From two observations as little as an hour apart, the west-to-east rotation is detectable. A typical amateur telescope is enough to show Mars' rotation, although digital photography with image enhancement makes it a lot easier to see what is there.

Astronomers of past generations, not least Percival Lowell in *Mars and its Canals*, Macmillan, 1906, have seriously, and notoriously, overestimated their ability to see what is on Mars (see Chap. 1). The reality is that you will see only the very largest features, and even these are a lot clearer when Mars is close to the Earth. Your view will be much as if you could just about make out the continents on Earth. The weather on both Earth and Mars can sometimes obscure the view, so be prepared to observe patiently.

A Martian day is called a "sol." In what follows, I will show you how to measure the duration of a sol. Strictly speaking, I am going to measure a sidereal sol.

The easiest feature to find on Mars is Syrtis Major (see Fig. 5.1. This was discovered in the early years of telescope astronomy by Christiaan Huyghens. No

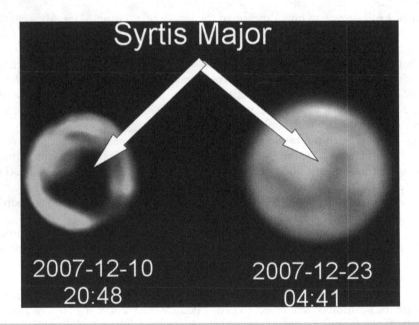

Fig. 5.1. At these two times, Mars presents approximately the same face to us. The two times are 1,065,285 s apart. The difference in seeing is striking: the photographs were taken and processed very similarly.

one really knew what it was until the planet was visited by space probes. It turned out to be a shield volcano. An account of Martian observation is given in *The Planet Mars: A History of Observation and Discovery* by William Sheehan.[32] It is interesting to note that the observers credited with the first great systematic survey of Mars, Beer and Maedler in 1828, used a 3.75-in. (95-mm) refracting telescope. By today's standards, that would be a very modest amateur instrument.

If you see the same face of Mars on two separate occasions, the number of rotations it has completed is a whole number. This is certainly not rocket science.

On the night of 22/23 December 2007, I took the sequence of pictures shown in Fig. 5.2. In these, the rotation of Syrtis Major is very evident. They are shown in Fig. 5.2.

At these two times, Mars presents approximately the same face to us. The two times are 1,065,285 s apart. The difference in seeing is striking: the photographs were taken and processed very similarly.

The North Polar ice cap is visible at the top of these pictures. This enables us to estimate the direction of North, not desperately accurately, but accurately enough as it will turn out.

If we take the line of longitude directly in front of the Martian axis to be zero, we can calculate the longitude by the method shown in Fig. 5.3.

Fig. 5.2. A sequence of photographs of Mars over a 4½-h period in the early hours of 23 December 2007. The rotation of Syrtis Major is very evident. It is also noteworthy that the seeing was not constant through the night. In addition, the color sensitivity of my webcam seems to be inconsistent. Some of this may have been my technique. Between 01:08 and 01:56 and between 02:56 and 03:41, I reset the camera to take pictures of Saturn and Titan, which are of course much fainter than Mars. Perhaps I could have done a better job of resetting the camera afterward.

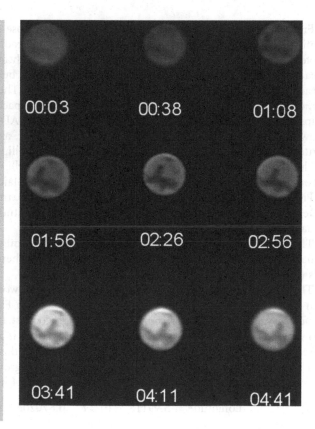

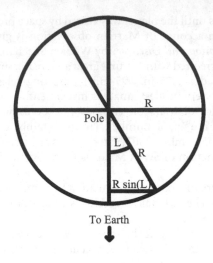

Fig. 5.3. Showing how to calculate the longitude of a point on Mars. The view is looking down onto the Martian North Pole. The Longitude is the angle L. By elementary trigonometry the distance from the Zero Longitude line to the point where of the ellipse crosses the Martian equator is $R \sin(L)$, since the ratio $R \sin(L)/R = \sin(L)$. Remember that the sine of an angle in a right-angle triangle is defined as "opposite over hypotenuse."

From Figs. 5.3 and 5.4, the longitude is the angle whose sine is the ratio of the semi-minor axis of the ellipse shown in Fig. 5.4 to the radius of Mars (or, if you prefer, the ratio of the minor axis to the diameter). I actually averaged the nine radii I measured. Figure 5.5 shows examples of this method being put into practice, measuring the ellipse of Fig. 5.4 on actual photographs. The distances in "inches" came from a computer-aided graphics (CAD) software package used to draw the ellipses and the circles outlining Mars. The tools in the CAD software were used to draw a circle around the planet to measure its diameter, a line through both the center of this circle and through the North Pole, and an ellipse, concentric with the circle, whose major axis goes from pole to pole. The ellipse is also made to go through the center of the northernmost point on Syrtis Major.

Figure 5.6 shows a plot of longitude of Syrtis Major vs. time generated from the nine photographs shown in Fig. 5.2. The units of longitude are radians. I could equally easily have used degrees. It does not matter.

The black line in Fig. 5.6 is a straight line fitted through these points, whose equation is shown in the caption at the top of the graph, where y is the longitude and x is the time.

The other lines show what the longitude change vs. time would look like if I assume that the number of sols between the two photos shown in Fig. 5.1 is 11, 12, and 13 respectively. If I assume 11 sols, the longitude changes too fast. If I assume 13, it changes too slowly. If I assume 12 sols, the rate of longitude change is almost exactly what was observed. I conclude that there were indeed 12 sols during this time interval.

The straight line fit, which assumes that longitude grows at 6.99118×10^{-5} radians per second, corresponds to a rotation period of 24 h 57.9 min:

$$\text{Longitude} = 6.99118 \times 10^{-5}t - 0.8702030 \text{ radians}, \qquad (5.1)$$

from Fig. 5.6, where t is in seconds.

Fig. 5.4. The great Circle of Longitude in Fig. 5.3 appears as a *straight line* when viewed from above the poles. When viewed from above the equator, it is an ellipse, as shown here. The Earth is roughly above the Martian Equator, but only roughly. In December 2007 we can see the North Pole more clearly than the South Pole. The semi-minor axis of the apparent ellipse shown has length $R \sin(L)$, where R is the radius of the planet.

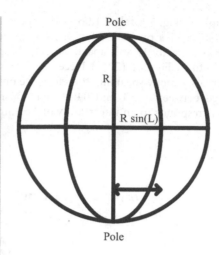

Fig. 5.5. Showing how to estimate the longitudes of features on Mars relative to a meridian which is directly in front of the planet's axis as we see it on Earth. Great circles through other lines of longitudes are seen from Earth as ellipses. The ellipses have been chosen to go through the center of the northern limit of Syrtis major, as accurately as I can estimate it. The distances in "inches" come from a computer-aided graphics (CAD) software package used to draw the ellipses and circles outlining Mars.

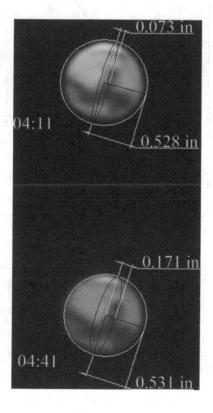

The two photographs of Fig. 5.1, taken 1,065,285 s apart, imply a rotation period of 1,065,285/12 = 88,773.75 s or 24 h 39 min 34 s, if we assume that the two photographs were taken exactly 12 sols apart. In fact, the photo in Fig. 5.2, from which the second photo in Fig. 5.1 (22/23 December) was chosen, were taken at roughly 30-min intervals. Strictly speaking, therefore, we only know that the end of the 12th sol occurred between 04:11 and 04:41. This 30-min uncertainty in the duration of 12 sols corresponds to an uncertainty of 2.5 min per sol.

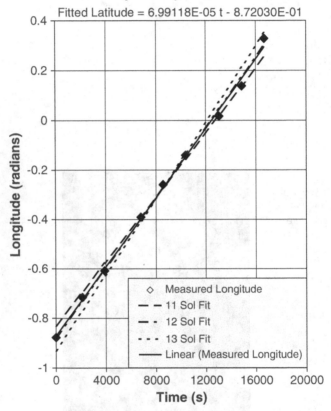

Fig. 5.6. Plot of longitude of Syrtis Major vs. time generated from the nine photographs shown in Fig. 3. The units of longitude are radians. I could equally easily have used degrees. It does not matter. The *solid line* is the best fit straight line fitted through these points, whose equation is shown in the caption at the top of the graph, where *t* is the time in seconds. The other *lines* show what the longitude change vs. time would look like if I assume that the number of Martian days ("sols") between the two photos shown in Fig. 5.1 is 11, 12, or 13, respectively.

Thus

$$1 \, \text{sol} = 24 \, \text{h} \, 39 \, \text{min} \, 34 \, \text{s} \pm 2 \, \text{min} \, 30 \, \text{s}. \tag{5.2}$$

Can we do better? In Table 5.1, I show the results of a series of repeat measurements, as well as the original measurements. For the later measurements, unfortunately, there is a joker in the deck. The further Mars is from opposition, the further it is from full. The further after January 25, 2008, the more difficult it was to make my measurements.

From Table 5.1, it is noticeable that if the December 10th observation is one of the pair, the rotation periods are somewhat scattered. If that observation is rejected, the mean and standard deviations of the remaining three pairs of observations give

$$1 \, \text{sol} = 24 \, \text{h} \, 37 \, \text{min} \, 36 \pm 5 \, \text{s}. \tag{5.3}$$

The published value,[33]

$$1 \, \text{sol} = 24 \, \text{h} \, 37 \, \text{min} \, 26 \, \text{s}, \tag{5.4}$$

lies almost within this range, as does the Wikipedia (http://en.wikipedia.org/wiki/Mars) value at the time of writing,

$$1 \, \text{sol} = 24 \, \text{h} \, 37 \, \text{min} \, 23 \, \text{s}. \tag{5.5}$$

The reason why the measurement taken over several days is so much more accurate than the one made purely on the night of 22/23 December is simple. Counting, in this case counting days, is an exact process. It does not involve any of the kind of guesswork involved in estimating longitudes of Syrtis major in Figs. 5.2 and 5.5. Furthermore, the uncertainty in the measurement of the longitudes is shared over several sols.

The difference between (5.3) and (5.5) is a remarkable 0.01%.

My measurements were made with an 8-in. telescope, a webcam, and the clock in my laptop. It is well within the scope of a typical amateur astronomer to make accurate measurements of the rotation of Mars. The same method can be used for Jupiter, because its great red spot makes an easy target. Other planets are not so easy because of the difficulty of seeing their surfaces. The very best amateur astrophotographers can just about observe the rotation of Jupiter's moons (see e.g., http://www.damianpeach.com/images/articles/peachfeature/%5B80-81%5DDamian%20Peach%20dec06.pdf) although this requires much better instruments than most of us can afford.

The Celestial Sphere

Figure 5.7 shows what the celestial sphere is. It is an imaginary sphere much bigger than the Earth, through which we look at the heavenly bodies. It co-moves with the Earth, except that it does not rotate about the Earth's axis. Hence the stars, which are so far away that parallax effects are extremely hard to detect, each appear to occupy a very nearly constant location with respect to the celestial sphere. We give these locations a kind of polar coordinates. The units of these in the north–south direction

Table 5.1 Data from Observations of Syrtis Major, December and January 2007/8

Date	39426	39439	39459	39472		
Time	0.866666667	0.195138889	0.715277778	0.030556		
Days since 00:00 on Dec 10	0.866666667	13.19513889	33.71527778	46.03056		
Radius from CAD photo	3.727	3.586	2.972	3.553		
Semi minor axis of logitude ellipse from CAD photo	0.989	1.202	1.015	1.189		
Longitude of tip of Syrtis major	0.268578202	0.341809452	0.348534577	0.34123		
Pair of Observations	Dec 10–Dec 23	Dec 10–Jan 12	Dec 23–Jan 12	Dec 10–Jan 25	Jan 12–Jan 25	Dec 23–Jan 25
Change in Longitude (radians)	0.07323125	0.079956375	0.006725125	0.072652	−0.007304206	−0.000579081
Change in Longitude (rotations)	0.011655115	0.0127725452	0.001070337	0.011563	−0.0011625	−9.21637E-05
Number of rotations	11.98834488	31.98727455	19.99892966	43.92735	12.0011625	32.00009216
Time interval (sec)	1065180	2838120	1772940	3902160	1064040	2836980
Seconds /rotation	88851.29768	88726.53391	88651.74436	88832.13	88661.41092	88655.36966
Days	1	1	1	1	1	1
Hours	0	0	0	0	0	0
Minutes	40	38	37	40	37	37
Seconds	51	47	32	32	41	35

The top of the table refers to individual observations. The lower half of the table estimates the length of a sol from pairs of observations

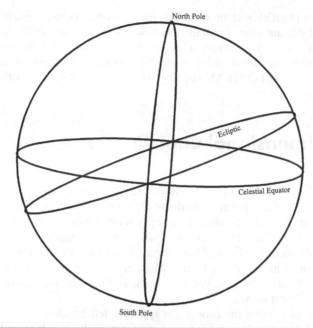

Fig. 5.7. The celestial coordinate system is conceptually simple. We imagine an enormous sphere of radius R, whose value does not matter so long is it is much bigger than that of the Earth. We imagine a sort of transparent sphere which has two poles above the Earth's poles, and an equator above that of the earth. This sphere does not rotate with the earth, but it does move so that its center is always at the center of the earth. The ecliptic is a circle with the same center and radius as the celestial sphere, along which the Sun appears to move throughout the year. It is known to be at an angle of about 23.4° to the celestial equator.

are simply degrees, just like the degrees of latitude on the Earth's surface. The celestial equator is at 0°; the celestial North Pole is at +90°; and the celestial South Pole is at −90°. These north-south angles are called *declinations*.

The units in the East–West Direction are called *right ascension*. This is an unhelpful, rather counterintuitive name, but we are stuck with it. Its units are equally irksome: right ascensions are measured in hours, minutes, and seconds from the point where the ecliptic crosses the celestial equator at the March equinox (see Fig. 5.7). They go from 0 to 24 h, increasing as you go eastward.

This zero point in right ascension is sometimes called the "First Point of Aries." Never mind that it is not even in the constellation of Aries, but is in fact in Pisces. Of course, due to the precession of the equinoxes, this position drifts on a timescale of centuries, but that is of no practical interest when measuring distances to planets, except that every half-century or so, the star catalogs get updated to allow for this. They quote coordinates relative to an "epoch" such as J2000, which effectively means "updated in the year 2000."

I do not intend to stick with these archaic units. Instead, I will normally quote right ascensions and declinations in radians, because it will make doing the trigonometry a whole lot easier. I use spreadsheets a lot for my calculations, and the most common spreadsheets use radians for their trigonometric functions like sines. So, if it comes to that, do languages like FORTRAN and C++ if you prefer to do your calculations by writing programs.

Three-Dimensional Coordinates

I wish to define a Cartesian coordinate system for the Celestial Sphere, to make it easier to transform into a coordinate system based on the ecliptic.

I am going to make x point toward the First Point of Aries, where the ecliptic crosses the celestial equator in March; and z point toward the North Celestial Pole. In Chap. 6, I will make the right ascension radians of Leo negative. This was done to make the graph come out the right way round, not as a mirror image. For this to happen, my y-axis has to point to the West, giving positive right ascensions (in radians) to the West of the First Point of Aries, and negative ones to its east. Leo is east of the First Point of Aries.

This in turn means that my coordinate system is left-handed.

To change right ascensions from hours, minutes, and seconds to radians, I used the formula

$$\text{Radians} = -\frac{2\pi(\text{hours} + (\text{minutes}/60) + (\text{seconds}/3{,}600))}{24}. \quad (5.6)$$

I then used (A.13)–(A.15) in the Appendix A to turn my right ascensions and declinations into (x, y, z) coordinates, remembering that $\theta = (\pi/2) - D$, where D is the declination.

Next, I want to know where the ecliptic is.

Let Now the Sun Go Down Along Its Path

First, I found one point on the ecliptic. I had a photo of Saturn and its neighboring stars taken within a few hours of Saturn's opposition, i.e., the time at which the Sun, the Earth, and the Saturn lie in a straight line (or at least they would if the orbits were exactly coplanar). The sun would therefore have been diametrically opposite to the Saturn (from the Earth).

In particular, if, at opposition, the x- and y- coordinates of Saturn on the celestial sphere are (x_{SO}, y_{SO}), then the x- and y- coordinates of the Sun on the celestial sphere at that time must be $(-x_{SO}, -y_{SO})$.

The other key piece of information is sunrise and sunset times. Most amateur astronomers, myself included, follow these times with some care because we all want to know when it will be dark. Therefore, at the very least, most of us will have a rough idea when the sun rises and sets at our own locations throughout the year.

It can pay to think carefully about what you know, because you might know more than you think. I do not apply this principle to my bank account. I prefer to be in denial about the size of the overdraft, but thinking about astronomy is a much less stressful pastime.

In particular, if you know when the sun rose and when it sets, you know that the solar midday on that day was exactly halfway between them. That gives you a time for the solar midday. It drifts from mean time by up to 15 min because the Earth's orbit is a little bit elliptical, not a perfect circle, and because the ecliptic is not parallel to the celestial equator, so you do need to know when the solar midday is.

You also know that the Earth rotates at a very nearly constant rate. If you are an astronomer, you know that the rotation period is called a sidereal day, and it is 23 h, 56.1 min or 0.99727 days.[33] All amateur astronomers experience the way in which a given star rises about 4 min earlier each day. Of course, the reason why the earth does not rotate in exactly 24 h is that, as it orbits the Sun, the orientation of the Sun relative to the stars keeps changing. The Earth actually completes 366¼ revolutions in a year.

From the rotation rate, my solar middays plus my Saturnian opposition, I was able to work out the orientation of the Sun at midday on every day of the year from 30 August 2007 to the same time in 2008. I chose this particular year because 30 August 2007 was the date when I began to track Mars. That gave me the (x, y) coordinates of the Sun on the celestial sphere at every Solar midday.

What about the z coordinates? The variation in z of the solar position is of course what gives us our seasons, our long nights in Winter and our long days in Summer. That is the key: if at any point on the earth it is sunrise or sunset, the sun is perpendicular to the zenith. This never happens at any other times at that location. During the day, the Sun is above the horizon, so the angle it makes with the zenith is acute. At night, it is below the horizon, so the angle with the zenith is obtuse.

I will now go through this calculation in detail.

φ Noon

During a 90-s photographic exposure starting at 00:08:07 GMT on 25 February 2008, I measured the right ascension and declination of Saturn near opposition to be 10 h 29 m 45 s and 11°5 m 9 s, respectively, or RA $= -2.7478$ radians and Dec $= 0.1935$ radians. The method I used will be described in Chap. 6.

I now convert these into my Cartesian coordinates using the formulae

$$x = r \cos(\varphi)\sin(\theta), \quad y = r \sin(\varphi)\sin(\theta), \quad \text{and } z = r \cos(\theta), \qquad (5.7)$$

where φ is the right ascension in radians, $\theta = ((\pi/2)-\text{declination})$ in radians, and r is the radius of our celestial sphere. These formulae are derived in Appendix A. The results are

$$x = r\cos(-2.7478)\sin\left(\left(\frac{\pi}{2}\right) - 0.1935\right) = -0.9062r \quad \text{and}$$

$$y = r\sin(-2.7478)\sin\left(\left(\frac{\pi}{2}\right) - 0.1935\right) = -0.3765r. \tag{5.8}$$

Hence the (x, y) coordinates of the Sun at this time must have been

$$(x, y) = (0.9062, 0.3765)r. \tag{5.9}$$

I need to declare an approximation here: I assumed the Sun's right ascension did not change in the 12 h between midnight and noon on 25 February. This introduces an error of half a part in 365.25, or about 0.14%. I will live with that.

Next, I choose the day on which I wish to know the (x, y) coordinates of the Sun. Since there will be many such days, this calculation is best done in a spreadsheet, such as Microsoft Excel or OpenOffice Calc.

I actually got my sunrise and sunset data from the US Naval Observatory's Web site http://aa.usno.navy.mil. This is by no means unpatriotic of me: their Almanac is compiled jointly with the British Navy. Was this cheating? Well, sort of. I use this site routinely, e.g., when I was tracking Venus and Jupiter over much of 2007, so I know that it is at the very least approximately correct, although I have not verified it by careful daily timing. It is not easy to spot sunrises and sunsets, especially the latter, because you have to look through a lot of atmosphere to see the horizon. At English latitudes, it cools down in the evening before sunset, so clouds often form around sunset; and you do not see it.

In fact, nowadays you can get a measure of these times in all weathers. The screen of my car's satellite navigation system changes its background color from black to white at sunrise, and back again at sunset. I know from experience that when it decides to change color, it corrects for position as I drive around.

Anyway, let us take the day on which I am writing this as my example: May 31, 2008. According to Uncle Sam, sunrise was 03:41 GMT and sunset was 20:12 GMT. They are both in the past: I am writing this on a laptop wired up to my webcam as I wait for the clouds to clear so I can photograph Saturn. The early sunrises mess up my sleep patterns, so I certainly know that they happen. The time of Solar noon was midway between these, 11:56. It is convenient to express this time as a fraction of the day: 0.4975 days. In Microsoft Excel, you only need to convert the time format in the data cell to a number format and this happens automatically.

It is 96 days from February 26 to May 31 in a leap year like 2008. 96 days is equal to

$$\frac{96}{0.99727} = 96.2628 \text{ sideral days}, \tag{5.10}$$

and therefore from noon GMT to noon GMT, the earth rotated 96.2628 times.

The 4-min difference between solar noon and noon GMT corresponds to -0.0024 sidereal days. The corresponding figure on February 25 was similarly calculated to be $+0.00836$ sidereal days: solar noon that day was at 12:11.

Now I know enough to work out how many times the Earth rotated between 25 February and 31 May: $96.2628 - 0.0024 - (+0.00836) = 96.252$ times. It has therefore changed its orientation relative to the celestial sphere by 0.252 revolutions, or

$$2\pi \times 0.252 = 1.584 \text{ radians}. \tag{5.11}$$

On 25 February, The Sun's right ascension was

$$\text{Saturn's RA@opposition} + \pi = -2.7478 + \pi \text{ radians}. \tag{5.12}$$

Therefore on 31 May it would have been

$$
\begin{aligned}
\varphi_{Noon} &= \text{Saturn's RA@opposition} + \pi - 1.584 \\
&= -2.7478 + \pi - 1.584 \text{ radians} \\
&= -1.190 \text{ radians}.
\end{aligned}
\tag{5.13}
$$

On the θ Front

To continue with my example, I need to tell you my latitude. It is 52°46 m 19 s North. (My longitude is 0°26 m 02 s East, but that is less relevant.) I got these data from the internet map http://www.multimap.com. The same Web site also works in the United States: it gives the coordinates of my wife's childhood home in Seattle as 47° 34 m 27 s North and 122° 18 m 23 s West. I cannot remember the Zip Code: all I did was type in the address.

Anyway, I have already told you the sunrise, noon, and sunset times. To get the Sun's declination from these requires a little bit of low cunning. Indeed, as stated earlier, at these times, the Sun is low: so low it is at right angles to the zenith.

I can exploit that fact! I know the coordinates of my positions on the Earth at these times relative to those at noon; or at least I can easily work them out. I then extrapolate a line from the Earth's center through my backyard out as far as I like. I then assume that the Sun is so far away that its direction is the same from any point on Earth; and I neglect the apparent motion of the Sun along the ecliptic during May 31. I imagine a line from the Earth's center toward the Sun. It does not have to go all the way to the Sun. It can go further, or less far. It does not matter; this vector will still be perpendicular to my zenith vector at sunrise and sunset.

Let us call these lines my zenith vector U (for "up") and the solar vector S. I show in the Appendix A, how the scalar or "dot" product of two vectors is equal to zero if the two vectors are at right angles to one another. I further show that the dot product is equal to $U_x S_x + U_y S_y + U_z S_z$, where U_x, U_y, and U_z are the x-, y- and z-components of U, and S_x, S_y, and S_z are the x-, y- and z-components of S. Therefore

$$U_x S_x + U_y S_y + U_z S_z = 0. \tag{5.14}$$

This gives me an equation to solve for S_z. I can do it, because in principle I know all the other components, even if I have not yet worked them out in detail.

I cannot translate my value of φ_{Noon} for May 31st into the x- and y-coordinates on the celestial sphere using (5.7), because I do not yet know the declination θs. However, I do know that the values are

$$S_x = r\cos(-1.190)\sin(\theta_s) \text{ and } S_y = r\sin(-1.190)\sin(\theta_s); \quad \text{or}$$
$$S_x = [r\sin(\theta_s)]\cos(-1.190) \text{ and } S_y = [r\sin(\theta_s)]\sin(-1.190). \tag{5.15}$$

I have taken the value of φNoon from (5.13). As long as I am careful to make my value of S_z be

$$S_z = [r\sin(\theta_s)]\cos(\theta_s) \tag{5.16}$$

I have not cheated. The distance along S is not r, the distance to the celestial sphere, but $r\sin(\theta_s)$.

What about little me here on Earth? Well, I know from the sidereal rate of terrestrial rotation and from the times of solar noon and of sunset that I have rotated around the Earth's axis by an amount determined as follows: The solar noon was at 11:56. Sunrise was at 03:41. I have made a slight correction because the almanac definition of sunrise is the appearance of the top of the Sun above the horizon. Very crudely, and a bit wildly, I have said that, from the well-known fact that the sun occupies half a degree of arc in the sky and from the fact that it does not rise vertically, the midpoint of sunrise was 1/720 of a day later, or at 03:44. That is 8 h 12 min, or 8.2 h, or $(8.2/24) \times 0.9972 = 34.17\%$ of a sidereal day before noon. This corresponds to a rotation of 2.152 radians.

At solar noon, my zenith vector will have the same right ascension as the Sun, though of course a different declination. So at sunrise I simply add my 2.152 radians to the right ascension of the noon solar vector, and I have the right ascension of my zenith vector. I add rather than subtract in moving backward from noon to sunrise, because my zenith vector moves eastward, and I set eastward φ to be negative when I converted to radians. I know the declination of my zenith vector: it is simply my latitude. In spherical polar coordinates, the zenith vector at sunrise was therefore

$$U = (R_{\text{birtrary}}, \phi_{\text{Noon}} + \psi_{\text{Noon-Sunrise}}, (\pi/2) - \lambda)$$
$$= (R_{\text{birtrary}}, \phi_{\text{Zenith@Sunrise}}, \theta_{\text{Zenith@Sunrise}}), \tag{5.17}$$

where $\psi_{\text{Noon-Sunrise}}$ is the Earth's rotation between sunrise and noon, λ is my latitude, and R_{bitrary} is the arbitrary (R-bitrary: geddit?) length of the zenith vector. All I care about with U is that it is at right angles to S at sunrise. I am not worried how long it is.

I now collect up what I have worked out and substitute (5.7), (5.15), (5.16), and (5.17) into (5.14). Then

$$U_x S_x + U_y S_y + U_z S_z = [r\sin\theta_s]R_{\text{bitrary}}$$

$$\times \begin{pmatrix} \cos(\varphi_s)\cos(\varphi_{\text{Zenith@Sunrise}})\sin((\pi/2) - \lambda) \\ +\sin(\varphi_s)\sin(\varphi_{\text{Zenith@Sunrise}})\sin((\pi/2) - \lambda) \\ +\cos(\theta_s)\cos((\pi/2) - \lambda \end{pmatrix}$$

$$= 0 \tag{5.18}$$

I can cancel the factor $[r\sin\theta_s]R_{\text{bitrary}}$ in (5.18). I can also simplify the latitude terms using $^{(A.2)}$ in the Appendix A: $\sin((\pi/2) - \lambda) = \cos(\lambda)$ and vice versa. Then

$$\left(\begin{array}{l} \cos(\varphi_s)\cos(\varphi_{\text{Zenith@Sunrise}})\cos(\lambda) \\ + \sin(\varphi_s)\sin(\varphi_{\text{Zenith@Sunrise}})\cos(\lambda) \\ + \cos(\varphi_s)\sin(\lambda) \end{array} \right) = 0. \tag{5.19}$$

The only unknown in (5.19) is $\cos(\theta s)$. You can solve it for this. I actually used a simplification based on knowing that

$$\cos(\varphi_s)\cos(\varphi_{\text{Zenith@Sunrise}}) + \sin(\varphi_s)\sin(\varphi_{\text{Zenith@Sunrise}})$$
$$= \cos(\varphi_s - \varphi_{\text{Zenith@Sunrise}}) \tag{5.20}$$

but if (5.20) looks like black magic to you, do not worry. I only mention it in case some wit accuses me of not noticing this and writes a bad review. You do not need to use (5.20); and its derivation is not the world's easiest.

Without (5.20), I plug the numbers I know for May 31 into (5.19), namely $\varphi_s = -1.190$, $\varphi_{\text{Zenith@Sunrise}} = 0.958$, and $\lambda = 0.650$, all in radians, and I obtain

$$\left(\begin{array}{l} 0.372 \times 0.576 \times 0.796 \\ +(-0.928) \times (0.818) \times 0.796 \\ + \cos(\theta_s) \times 0.605 \end{array} \right) = 0. \tag{5.21}$$

Whence

$$\theta = 1.178 \text{ radians} \tag{5.22}$$

As implied, I used a spreadsheet (Microsoft Excel) to do this calculation for the year August 30, 2007 through August 29, 2008. The results are shown in Fig. 5.8. They are shown as x-, y- and z-components of the vector S. I have compared them with the values downloaded from the US Naval Observatory. It is very hard, even when I look at the raw Excel graph on my computer screen, to see a difference between the Navy's values of the x- and y-components and mine. I can see a

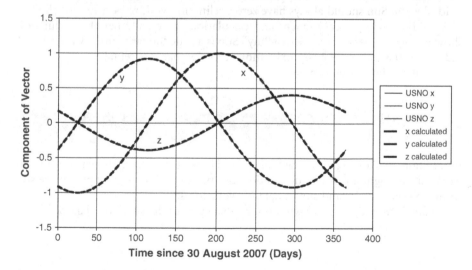

Fig. 5.8. Components of x, y, and z components of the vector S pointing to the sun.

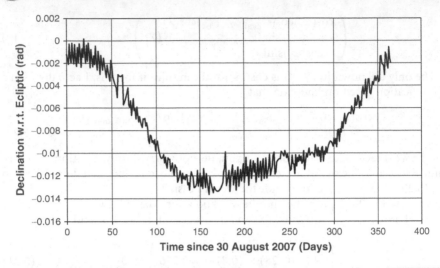

Fig. 5.9. The Declination of the Sun, according to my measurement and calculation, with respect to (w.r.t.) the ecliptic, assuming a tilt of 23.44°.

difference between the z-components. Mine are a little bit too far South. I have the Sun overhead half a degree too far South at both solstices.

Another way to check my calculation is to take the published value of the Earth's tilt (23.44°: according to Wikipedia's Earth article, 23.45° according to ref. 33, p. 57), and transform the components of my vector S so that the equator of my new coordinates is the ecliptic for a 23.44° tilt.

The method is given in the Appendix A, (A.18)-(A.26).

Ideally, the Sun should always have zero declination with respect to the ecliptic.

According to my measurement and calculation, it actually has the declination shown in Fig. 5.9. Again, we see that "my" Sun is a little too far South, by up to 0.014 radians, or 0.8°. Given the approximations I made, that is acceptable.

I could not have been contenter.

Conclusion

In addition to showing you how long a day is on Mars, I have shown you how to work out the right ascension, declination and position along the ecliptic of the sun from tables of sunrise and sunset data for your locality.

We are now ready to go to work on the distances to the superior planets.

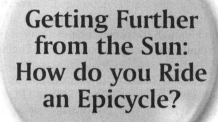

CHAPTER SIX

Getting Further from the Sun: How do you Ride an Epicycle?

I got the distance to Venus from a quick-hit measurement. For the superior planets, I was not saved from arduous labor by quick wit. I had to collect a lot of data and crunch many numbers. For both Saturn and Mars I collected more data than I absolutely needed. One reason for this was that I simultaneously read around to see if I could find a useable analysis technique. I simply did not know how much data I was going to need.

The other reason was that astronomy is my hobby, and I was simply having a bit of fun watching the apparent movement of the planets. I did not need to care very much about whether my science was efficient, unlike in the day job.

First, I will take you through my photographic techniques. I am not afraid to experiment, knowing that some ideas will bomb out – but I do keep things simple and stupid. Then I will show you how to extract the coordinates of a planet from a photo. My simplified model of the planet's orbit is that it is circular, so I will take you through the geometry of circular orbits, and show you how to work out what you would expect to see for such an orbit. Finally, I will compare such orbits with my photographic data, and show you that by assuming orbits to be circular, you can get a good distance to Saturn and a reasonable one for Mars.

I suppose you could call the case for using circular orbits the circular argument...

J.D. Clark, *Measure Solar System Objects and Their Movements for Yourself!*,
Patrick Moore's Practical Astronomy Series,
DOI: 10.1007/978-0-387-89561-1_6, © Springer Science + Business Media, LLC 2009

Photography

Not the least of the talents of the early astronomers such as Galileo was their ability to draw accurately. In some circles, such as the Internet-based Society for Popular Astronomy, to which I belong, this is still encouraged. I have the greatest respect for people who can do this, not least because I cannot draw at all well.

If your drawing is good enough to record the positions of planets relative to the stars, you could most certainly track the planets that way. That is how people used to do it.

However, I am reduced by lack of talent to photography for all my imaging.

For astronomical purposes I use a couple of Philips SPC900NC^{TM} webcams. I bought a second because I found that continually swapping the lens and the telescope adapter, with a UV/IR filter as a dust cover, was resulting in a lot of dust on the chip.

I hope by now you have latched onto the idea that to make most measurements, you do not need the kind of photographic quality that will get you published in *Sky and Telescope*. What you do need is fairly regular pictures, whether the seeing is good or lousy. The people who get shots published in magazines do not show you the photos they canned because they had to wait days if not weeks for good seeing. We do not have that luxury here.

My first method of photographing the planets was to place the webcam on a camera tripod, and shoot a movie of the sky. The field of view was about enough for the constellation of Leo, or for the two constellations of Gemini and Auriga. To deal with the tracking, I simply let my stacking software Registax^{TM} take care of the apparent movement of the sky. How much exposure was needed depended on the conditions. Ninety seconds at five frames per second was as much as I needed. I tried a 5-min movie, but there was too much rotation for Registax to cope with. On really clear nights I could have got away with less, but by the time I had reprogrammed my video capture software – K3CCD Tools^{TM} – the movie was already shot.

The method really only worked when there was a bright planet in the picture. Most nights, the stars twinkled too much for Registax to track them. It can track a handful of really bright stars such as Capella (α Aurigae). I tried photographing constellations this way. Orion was pretty easy, and I got a nice picture, but Ursa Major was a complete failure. There was nothing bright enough to track.

If there was a trackable object in the shot, then on clear nights I could certainly see more stars in the photo than with the naked eye. This is hardly a severe test of the technique: in a suburban backyard with a laptop computer next to me, I was not exactly dark adjusted.

The worst nights were the ones with slight fog that you could barely see. The orange glow of the street lights reflects depressingly well from the water droplets. I suppose I could have experimented with light pollution filters, but it was not really necessary. I got all the photos I wanted; and rejected very few as too poor to make measurements from.

Also, this camera comes with a glass lens, which will keep out the UV and the IR to which the chip is sensitive.

Fig. 6.1. One of the better photos taken with a webcam. You can see plenty despite the presence of the Moon. The Moon was only a problem on misty nights. The *blossom* on the tree branch shows how much street light I have to live with. This vegetation is not sharp because Registax was tracking Mars, not the tree. I would like to show you a poor but useable photo, but I doubt if it would reproduce in a book. Suffice it to say that if you can see ten stars well enough to find their centers on the picture, you have a useable photo. On this one, on my PC screen, I can see 28 stars.

As time went on, I realized that one of the difficulties with this technique is that it is not very good on partly cloudy nights. I was tending to take few pictures because there were not enough completely clear nights. In the event, this proved not to be a serious problem; I had enough data to make good measurements (Fig. 6.1).

At one level, this photography was easy. With Registax, it took little skill to obtain good pictures. There was a bit of an art to setting the exposure level in the webcam driver, so that the sky was dark but the stars were still visible, but not much of a one. If the picture was no good, I just tried again.

Eventually, however, I became dissatisfied with this technique, mostly because it was hard to triangulate accurately to get good right ascension and declination coordinates for the planets. I discuss this point below.

Another method, which I have not tried, is to attach a single-lens reflex camera with a telephoto lens to a guided telescope mount and take a long exposure photo. I am sure that this tried-and-tested method will work, but I cannot tell you from experience how good it is at tracking planets.

As an experiment, I had a go at placing the webcam behind a 26-mm eyepiece on my 8 in. *f*/6 telescope. This technique is known in the photography books as *afocal coupling*,[34] although I have also seen it called *eyepiece projection*. Learning to do this was fiddly, but rewarding (Fig. 6.2).

To hold the webcam onto the eyepiece, I bought a digital camera holder from a nearby firm called Scopes'n'Skies™, who share premises with a firm called Astro Engineering™ that makes little mechanical add-ons for astronomy. The webcam did not fit, but, fortunately, they had lots of spare Astro Engineering bits and sold me the extra piece I needed to make the assembly work. Although they were, as always, courteous and helpful, I did detect some skepticism that this arrangement would work. I was not sure either, but was willing to risk £25 ($50) in the attempt.

I need not have worried. It works just fine on my 6-in. and 8-in. reflectors. I only found out about the 6-in. instrument because of an emergency. One Friday night, the electronic dual axis drive controller on my Sky-Watcher EQ5™ mount began to sizzle and emit burning smells. It never worked again.

"Er, that'll take six weeks to obtain, Sir."

"But it can't. I have a book to write and a publisher's deadline. I really need to resume these observations."

To give James at Scopes'n'Skies his due, he understood and got me a new controller in 4 days. Meanwhile, I had to use the 6-in. machine, which has its own mount and controller.

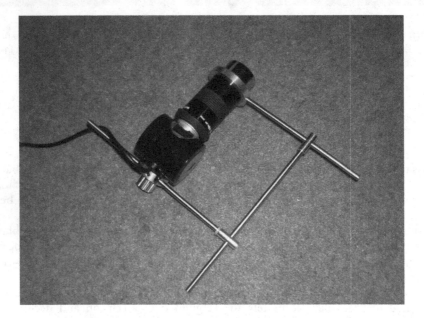

Fig. 6.2. Attaching the webcam to an eyepiece in afocal coupling mode. The holder was supplied by a firm about an hour from here called Scopes'n'Skies, who are on the same premises as a firm called Astro Engineering that makes little mechanical widgets for astronomy. There is an irritating LED on top of the camera, which I covered with electrical tape.

With the 8-in. beast, I found that a 5 min exposure at five frames per second (fps) gave better results than a 90-s exposure. After stacking, there was much less pixel noise in the pictures. When Saturn was still high in the sky after dark, on some nights I could photograph stars as dim as Magnitude +12.5. Most nights I could capture Magnitude +10.5 ones.

The tricky bit was getting good alignment between the axis of the webcam lens and the axis of the eyepiece. Once I had achieved this, I tightened the grub screws so darn tight I wonder if I will ever get them undone. The one screw you cannot overtighten on those webcams is the one that goes into the thread at the base of the camera. If you do, you will detach the plastic mount holding the metal thread. Fortunately, it can be superglued back, but this is not ideal. Superglue is not very waterproof; and in my part of the world you do get dew.

The other, much more major, irritant about Philips webcams for astrophotography is that there is a bright white LED just above the lens. Even when covered with black electrical tape, this LED does cause stray light. No doubt by writing your own software driver, you could suppress this, but why do not they give you a radio button on the supplied driver to turn off this LED?

Astrometry: Getting Data from Photographs

I had to solve the problem of how to work out the right ascension and declination of a planet on a photograph. The method I am going to show you will only work if the star field can be approximated by a Mercator projection – you know, like on maps of the Earth where the north–south lines are always parallel and vertical (Fig. 6.3).

This approximation requires two preconditions. First, it is not good at all near the celestial poles – not that there are any Solar System planets there. Second, it only works over small areas of the sky. Even over the sickle of Leo, it is a bit crude. Keeping within the range from α Leonis to δ Leonis, or 0.20 to 0.35 radians of declination, gives acceptable results. Large ranges of right ascension do not ruin the approximation: it is large declination changes that require a better approximation (Fig. 6.4).

In Fig. 6.5, a Mercator projection is shown of two stars A and B, and a planet U, whose location is unknown.

It is worth noting that the names of A and B have been chosen so that the sequence A, B, and U is clockwise. Also, U is below the line AB. We will look at the case where U is above AB later. In Fig. 6.5, several angles have been worked out. I am now going to work out the tangent of the gradient of the lines AU and BU. The coordinates of A and B are (a_R, a_D) and (b_R, b_D), where the subscripts R and D refer to right ascension and declination, respectively. I am assuming that these are known. You can easily find them on the Internet using the usual search engines. Always use at least two sources of data to minimize the risk of misprints. At the time of writing, there is a Web site called http://www.sky-map.org with a very comprehensive, almost infinitely zoomable, sky map. It contains stars in the Henry Draper, Tycho, and US Naval Observatory catalogs. I cross-checked the stars I used against the original catalogs,

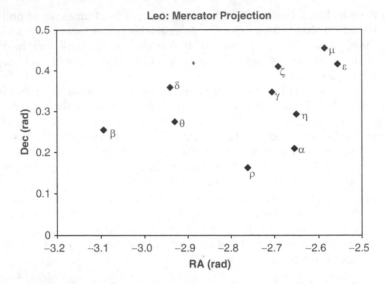

Fig. 6.3. A Mercator projection of the brightest stars in Leo. From my light polluted backyard, these are pretty much the only stars I can see. I have included ρ Leonis, which I can barely see, because it is on the ecliptic; and can be used to benchmark the movements of nearby planets. The units of right ascension and declination I have used are not the usual ones of a 24-h clock one way and ±90° the other. Instead, I have converted them to radians (see previous chapter). There are 2π radians in either 360° or 24 h. I also had to make the radians negative or I would have gotten a mirror image of Leo. Using radians will make subsequent calculations a little bit easier: software, especially spreadsheets like Microsoft Excel™, use them for trigonometry.

which are also on the Internet, and found one error, which I think must have been a misnamed star.

The difference in declension divided by difference in right ascension of a pair of stars is called its *gradient* in mathematical jargon. This gradient is also equal to the tangent of the angle the line makes with the right-ascension axis. In other words,

$$\frac{u_D - a_D}{a_R - u_R} = \tan(A - H);$$

$$\frac{b_D - u_D}{b_R - u_R} = \tan(B + H).$$

(6.1)

If you have half-forgotten what a tangent is, check out the Appendix A. Before proceeding to solve (6.1), I am going to remind you of the proof that the gradient of the angle is a tangent. This is proved in Fig. 6.6.

There is another hoop we have to jump through. What if the plane is above the line AB? In Fig. 6.7 I show that it does not matter so long as the stars A and B are named so that the sequence A, B, and U remains clockwise. Finally, I should point out that (6.1) break down if either of the denominators is zero or if the line AB is vertical. Avoid using reference stars that make this happen.

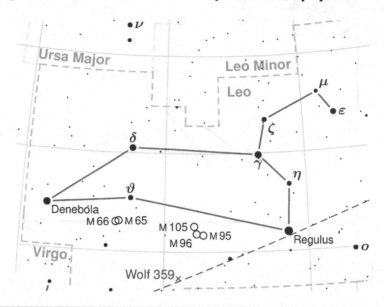

Fig. 6.4. A more accurate projection of the brightest stars in Leo. The *lines* of right ascension are not parallel. The *light dotted lines* are the boundaries of Leo: the darker one that passes by Regulus is the ecliptic. (Source: http://en.wikipedia.org/wiki/Image: Leo_constellation_map.png, where there is a GNU free document license granted.)

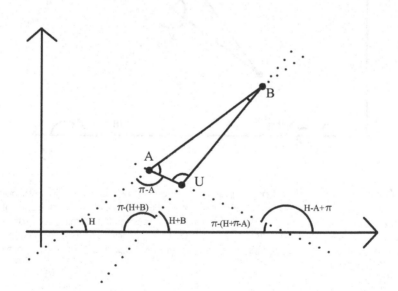

Fig. 6.5. Two stars A and B and a planet U. Some trigonometric angles have been worked out. Repeated use has been made of the rule that the angles of a triangle add up to 180° or π radians.

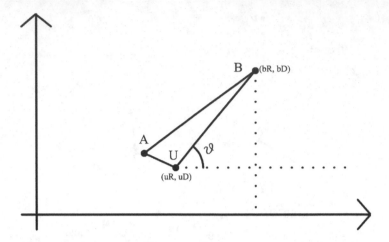

Fig. 6.6. Two stars A and B and a planet U. The tangent of θ is opposite divided by adjacent or $(b_D - u_D)$ divided by $(b_R - u_R)$.

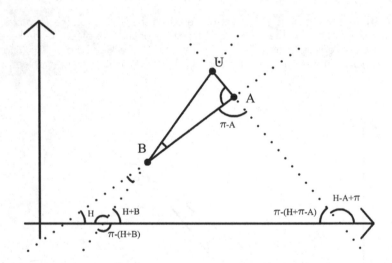

Fig. 6.7. Two stars A and B and a planet U, which is now above the line AB. Some trigonometric angles have been worked out. Repeated use has been made of the rule that the angles of a triangle add up to 180° or π radians. The angles governing the gradients of the lines AU and BU are the same is in Fig. 6.5. This only happens if I insist on naming stars A and B such that the sequence A, B, and U remains clockwise. Hence (6.1) are still valid in this case.

I am going to tell you what the solutions for (u_R, u_D) are, and then prove that they are correct. The as-yet-unproved solutions are (6.2):

$$u_R = \frac{a_D - b_D + [a_R \tan(A - H)] + [b_R \tan(B + H)]}{\tan(A - H) + \tan(B + H)};$$

$$u_D = \frac{[a_D \tan(B + H)] - [\tan(A - H)\{(b_D + [(a_R - b_R)\tan(B + H)])\}]}{\tan(A - H) + \tan(B + H)}.$$

(6.2)

To prove them, let me substitute them into (6.1). You then get

$$\frac{\frac{[a_D\tan(B+H)]+[\tan(A-H)]\{b_D+[(a_R-b_R)\tan(B+H)]\}}{\tan(A-H)+\tan(B+H)}-a_D}{a_R-\frac{a_D-b_D+[a_R\tan(A-H)]+[b_R\tan(B+H)]}{\tan(A-H)+\tan(B+H)}}=\tan(A-H)\ \text{and}$$

$$\frac{b_D-\frac{[a_D\tan(B+H)]+[\tan(A-H)]\{b_D+[(a_R-b_R)\tan(B+H)]\}}{\tan(A-H)+\tan(B+H)}}{b_R-\frac{a_D-b_D+[a_R\tan(A-H)]+[b_R\tan(B+H)]}{\tan(A-H)+\tan(B+H)}}=\tan(B+H).$$

(6.3)

What a mess! The quickest way to untangle this lot is to find common denominators. The obvious one is $\tan(A-H)+\tan(B+H)$. I then end up with a huge, messy pair of equations, which I am going to expand right out. All I am doing here is high school algebra – just rather a lot of it.

$$\frac{\frac{[a_D\tan(B+H)]+[\tan(A-H)]\{b_D+[(a_R-b_R)\tan(B+H)]\}-[a_D\tan(A-H)]-[a_D\tan(B+H)]}{\tan(A-H)+\tan(B+H)}}{\frac{-a_D+b_D-[a_R\tan(A-H)]-[b_R\tan(B+H)]+[a_R\tan(A-H)]+[a_R\tan(B+H)]}{\tan(A-H)+\tan(B+H)}}=\tan(A-H);$$

$$\frac{\frac{[b_D\tan(A-H)]+[b_D\tan(B+H)]-[a_D\tan(B+H)]-[\tan(A-H)]\{b_D+[(a_R-b_R)\tan(B+H)]\}}{\tan(A-H)+\tan(B+H)}}{\frac{[b_R\tan(A-H)]+[b_R\tan(B+H)]-a_D+b_D-[a_R\tan(A-H)]-[b_R\tan(B+H)]}{\tan(A-H)+\tan(B+H)}}=\tan(B+H).$$

(6.4)

I still need to expand out some brackets, and I also cancel some terms:

$$\frac{\left\{\begin{array}{c}[a_D\tan(B+H)+[\tan(A-H)]\{b_D+[(a_R-b_R)\tan(B+H)]\}\\-[a_D\tan(A-H)]-[a_D\tan(B+H)]\end{array}\right\}}{\left\{\begin{array}{c}-a_D+b_D-[a_R\tan(A-H)]-[b_R\tan(B+H)]+[a_R\tan(A-H)]\\+[a_R\tan(B+H)]\end{array}\right\}}=\tan(A-H);$$

$$\frac{\left\{\begin{array}{c}[b_D\tan(A-H)]+[b_D\tan(B+H)]-[a_D\tan(B+H)]\\-[\tan(A-H)]\{b_D+[(a_R-b_R)\tan(B+H)]\}\end{array}\right\}}{\left\{\begin{array}{c}[b_R\tan(A-H)]+[b_R\tan(B+H)]-a_D+b_D-[a_R\tan(A-H)]\\-[b_R\tan(B+H)]\end{array}\right\}}=\tan(B+H).$$

(6.5)

There is still more bracket expanding to do:

$$\frac{\left\{\begin{array}{c}[a_D\tan(B+H)]+[b_D\tan(A-H)]+[a_R\tan(B+H)\tan(A-H)]\\-[b_R\tan(B+H)\tan(A-H)]-[a_D\tan(A-H)]-[a_D\tan(B+H)]\end{array}\right\}}{-a_D+b_D-b_R\tan(B+H)+[a_R\tan(B+H)]}=\tan(A-H);$$

$$\frac{\left\{\begin{array}{c}[b_D\tan(A-H)]+[b_D\tan(B+H)]-[a_D\tan(B+H)]-[b_D\tan(A-H)]\\-[a_R\tan(A-H)\tan(B+H)]+[b_R\tan(A-H)\tan(B+H)]\end{array}\right\}}{[b_R\tan(A-H)]-a_D+b_D-[a_R\tan(A-H)]}=\tan(B+H).$$

(6.6)

Now I can start to look for terms which cancel. The result is

$$\frac{\begin{array}{c}[b_D\tan(A-H)]+[a_R\tan(B+H)\tan(A-H)]\\-[b_R\tan(B+H)\tan(A-H)]-[a_D\tan(A-H)]\end{array}}{b_D+[a_R\tan(B+H)]-[b_R\tan(B+H)]-a_D}=\tan(A-H);$$

$$\frac{\begin{array}{c}[b_D\tan(B+H)-[a_D\tan(B+H)]-[a_R\tan(A-H)\tan(B+H)]\\+[b_R\tan(A-H)\tan(B+H)]\end{array}}{b_D-a_D-[a_R\tan(A-H)]+[b_R\tan(A-H)]}=\tan(B+H).$$

(6.7)

If we reorder some of the terms, it will become clear that there is a great big cancellation we could do

$$\frac{\{b_D + [a_R \tan(B+H)] - [b_R \tan(B+H)] - a_D\} \tan(A-H)}{b_D + [a_R \tan(B+H)] - [b_R \tan(B+H)] - a_D} = \tan(A-H);$$

$$\frac{\{b_D - a_D - [a_R \tan(A-H)] + [b_R \tan(A-H)]\} \tan(B+H)}{b_D - a_D - [a_R \tan(A-H)] + [b_R \tan(A-H)]} = \tan(B+H).$$

$$(6.8)$$

Aha, we are getting there. Most terms now cancel, leaving

$$\tan(A-H) = \tan(A-H)$$
$$\tan(B+H) = \tan(B+H)$$

$$(6.9)$$

Equations (6.2) did not look very obviously true, but they are equivalent to (6.9). Not even those high priced lawyers who get rich crooks acquitted could argue that (6.9) are false.

With my webcam in afocal coupling mode, I could see a field of about one degree square. Over an area this small, the Mercator approximation is excellent. Doing ten triangulations would yield the same coordinates to within parts per thousand. The limitation on the technique was that it is not easy to determine the location of the center of Saturn when it is overexposed, only occupies a few pixels, and is sometimes a funny shape because of satellites which are not quite separately resolved.

I did not find this triangulation method in a book: my wife, daughter, and I discussed how to devise a simple method on a long car journey one day. We had plenty of time to kick ideas around, and tried writing a couple of them down when we got home.

I asked a professional astronomer friend, who once wrote a book on statistics for astronomers, why I did not find this apparently simple method in the textbooks. His answer was brief and to the point: professionals use mathematics to correct for optical distortion, whereas I have not.

How much does this matter? My wide-angle photos certainly suffered a lot of distortion. If I took overlapping photos and used Microsoft Digital Imaging Suite^{TM}, a cheaper version of Photoshop^{TM}, to overlay the images, they simply did not quite overlay. Compared to the Mercator-distortion error, this was small beer, so I did not get excited about it.

When I began using a telescope to track Saturn, I discovered something interesting. If a star, such as HD 89364 in Fig. 6.8, was toward the edge of the field of view, when the avi frames were stacked, optical distortion would sometimes make the star appear to be an elliptical blob, not a circular one. In other words, the star did not quite stack in the same place. If that happened, I simply did not triangulate against this star.

So, once again, digital photography and frame stacking have changed the rules of the astronomy game for the better. On a single-exposure 35-mm film shot, you would have to allow for distortion. On stacked digital photos, you can spot it and ignore it.

Ignoring a few stars is not that big a deal. If you can see n undistorted stars on the photo, each one can be paired with $(n-1)$ other undistorted stars for triangulation.

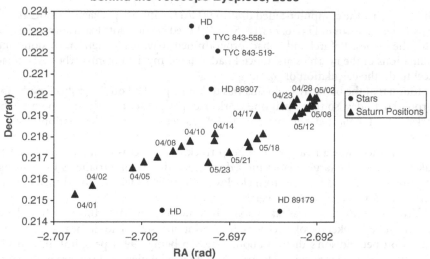

Fig. 6.8. Toward the end of the project I began to use a telescope to track Saturn, as opposed to a stand-alone webcam. This gave much more precise results, because I was no longer trying to impose a Mercator projection onto the whole of Leo. Dates are given in American format: 04/05 is April 5th. Some dates are omitted for clarity, especially near the turnaround from westward to eastward motion, when there was little apparent day-to-day motion. If I could wind the clock back, I would have used this tracking method for the entire project, although with fast-moving Mars, it would have been quite a challenge to identify all the nearby stars from a catalog. The gaps in the data give you a good idea of when it was cloudy all night. Once I had learned the technique, I needed about half an hour of intermittently clear sky to obtain a picture. I would shoot avi movie for 5 min if the clouds let me, which they usually did. Once I realized I could detect daily motion, even close to the stationary points, I began to take a lot more photos. Whether I would have had the patience to keep this up for 200 days I do not know. At the outset of the project I had no idea what I would detect. I simply had a go to see what I could see. I did not even know if I would detect any motion in Saturn. Mars is a different kettle of fish: you can see its day-to-day movement with the naked eye.

That makes $n(n-1)$ pairs. Oops! No, it does not. I have counted each pair twice. So the actual number of pairs of stars is $(1/2)n(n-1)$. It there are eight undistorted stars in the photo, there are $(1/2) \times 8 \times 7 = 28$ pairs of stars. That is a lot more than you need. Ten is plenty to check the reproducibility of the triangulation. Five good stars in the picture will give you $(1/2) \times 5 \times 4 = 10$ pairs of stars. By the way, you can make your friends do quite a good double-take by asserting after a couple of drinks that you have ten pairs of fingers on each hand.

I exploited this fact on my Mars photos to get around the poor Mercator approximation. Fortunately for me there were several stars in Gemini and Taurus with very similar declinations to Mars. I could usually find five of them to triangulate from. That was more by luck than judgment: it was not true for Saturn in Leo.

Actually Doing the Triangulation

I did this using the computer-aided drafting (CAD) software package TurboCAD™. Steps 1–4 are shown in Figs. 6.9–6.12. Step 5: I used a Microsoft Excel spreadsheet to enter the angles. Beforehand, I had programmed it with the right ascensions and declinations of the nearby stars. Once I had entered my data from TurboCAD, I had Excel to do the calculation of (6.2).

I found that the best way to do this was to have a pre-laid out worksheet, and copy it from day to day, so that I had one worksheet per photo. I used dates to name the worksheets. I collected the measured right ascensions and declinations into a summary worksheet, from which I could plot graphs. I found that copying each day's worksheet was an easy way to keep track of the occasional changes I had to make whenever the target planet moved so much that in one triangles I was using, the order of A, B, and U went from clockwise to anticlockwise. I would then have to switch star A to star B and vice versa.

I have to confess that I was taken aback by how much work all this was. It took me two or three weekends of pretty well continuous slog. Did it nearly drive me crazy? You bet! Edison's dictum about research being "99% perspiration and 1% inspiration" proved yet again to be true. You do need patience. If you are not used to CAD software, you will need to persist a little bit to become fluent with it. I always find this when I try a new CAD package: I am all fingers and thumbs for a while.

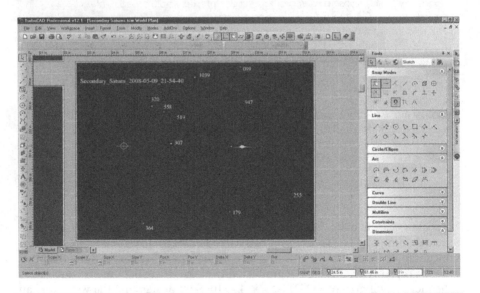

Fig. 6.9. Step 1: Import the png image from Registax into the CAD package, and identify the stars. I used abbreviated names: 320 for HD89320, 558 for TYC-843-558-1, and so on. My identifications came from the online catalog http://www.sky-map.org, which I checked against the online copies of the Tycho and US Naval Observatory catalogs. Links to these are given in sky-map.org.

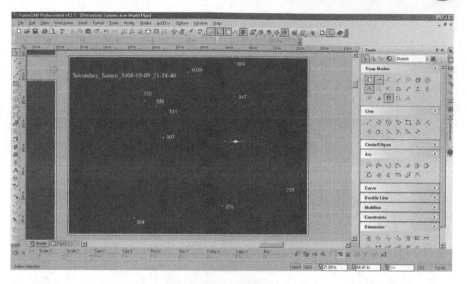

Fig. 6.10. Step 2: with the "snap" feature turned off, draw *circles* centered on each star. You will need to zoom in to each one to do this accurately. It may take several attempts to get a satisfactory result, especially with dim stars, but you will improve considerably with practice. Saturn presents a particular challenge which Mars does not: it is elliptical and occupies maybe 20 pixels in a shot like this. Sometimes this ellipse is distorted by satellites which cannot be separately resolved. I never found a really satisfactory answer to this. Drawing an ellipse and moving it around is about as good as it gets. Incidentally, in this picture, Titan is to the left of Saturn; and over several nights I was able to identify the object to the upper right of Saturn as Iapetus because it followed Saturn, not the Stars.

I also made the classic data analysis mistake of collecting a great pile of data before analyzing it. This was for the simple reason that, at the start of the project, I did not know how to analyze the data. In parallel with the photography, I was doing a great deal of reading to try to find a method. I got lots of ideas, but books written by celestial mechanics professors for their graduate students are not a rich source of simple techniques for amateur astronomers. Also, in fairness to the professors, there really is not a one-size-fits-all method. You have to use different methods for different orbits, depending on how great the eccentricity is, and how inclined to the ecliptic the orbits are.[35] Thus a technique that works well for Pluto, whose orbit is highly eccentric and inclined at 17° to the ecliptic, may completely fail to find the orbit of Saturn, whose orbit has low eccentricity and is close to the ecliptic (Figs. 6.13–6.16).

Three-Dimensional Coordinates

I used three-dimensional coordinates as described in the Appendix A. If you are not familiar with three-dimensional coordinates, I suggest you read this part of the Appendix, or you may get mental appendicitis.

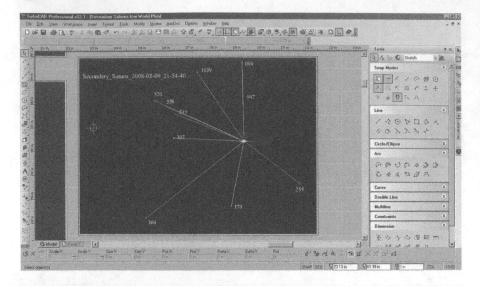

Fig. 6.11. Step 3: With the "snap" feature set to "center," draw *lines* from the centers of the *circles* where you think the stars are to the circle or ellipse you drew over the planet. This will place these lines with mathematical exactness to within the (very high) numerical accuracy of the software and the computer. Furthermore, you can do this very quickly. The hard work was done centering the circles on the stars.

I am going to make my *x*-axis point toward the First Point of Aries, where the ecliptic crosses the celestial equator in March; and *z* point toward the North Celestial Pole. You may recall that, in Fig. 6.3, I had the right ascension radians of Leo as negative. This was done to make the graph come out the right way round, not as a mirror image. For this to happen, my *y*-axis has to point to the West, giving positive right ascensions (in radians) to the West of the First Point of Aries, and negative ones to its east. Leo is east of the First Point of Aries.

This in turn means that my coordinate system is left-handed.

To change right ascensions from hours, minutes, and seconds to radians, I used the formula

$$\text{Radians} = -\frac{2\pi\left(\text{hours} + \left(\frac{\text{minutes}}{60}\right) + \left(\frac{\text{seconds}}{3,600}\right)\right)}{24}. \tag{6.10}$$

I then used (A.13) and (A.15) in the Appendix A to turn my right ascensions and declinations into (x, y, z) coordinates, remembering that $\theta = (\pi/2) - D$, where D is declination. Having done that, I used (A.24) and (A.26) in the Appendix A to obtain a set of (x', y', z') coordinates with the $x'y'$-plane being the ecliptic. This is easy to set up in Microsoft Excel or OpenOffice Calc. Using (A.16) and (A.17) in the Appendix A I could convert these back to polar coordinates. Again this is straightforward in a spreadsheet program.

From Figs. 6.17 to 6.20, you can see that the approximation that the planets orbit along the ecliptic is by no means perfect, but it is not a bad approximation.

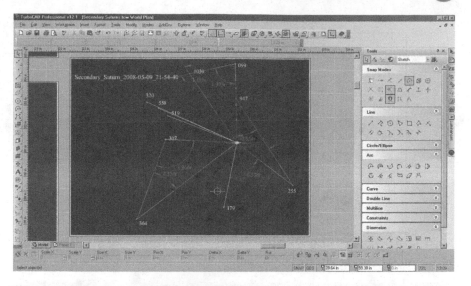

Fig. 6.12. Step 4: With the "snap" feature set to both "center" and "nearest object," draw *lines* between the centers of the chosen pairs of stars. Then use the dimensioning tools to measure the angles ABU and BAU per Figs. 6.5–6.7. I set the measurement units to be radians, and the precision to be three decimal places. For stars at very similar declinations to Mars, the angles were very shallow, and I got noticeably more reproducible positions for Mars if I used four decimal places. To prevent the image becoming too cluttered I put the lines AB (per Figs. 6.5–6.7) and the angles on different "layers" within the CAD file. In CAD jargon, "layers" are the software analog of different layers of tracing paper that draftsmen used to make their engineering drawings more intelligible in the not-so-good old days of drawing boards. I never found out what the limit to the number of layers in TurboCAD is: I have used over 80 in a single file in my day job.

If the planetary orbits all lay in exactly the same plane, then the planets would all move along the ecliptic, the imaginary line shown in Fig. 5.7. The eight known major planets orbit in approximately the same plane, but the superior planets only deviate from coplanarity by up to 2.5°.[36] They therefore appear to move very close to, but not quite on, the ecliptic. In the present analysis, we will ignore the slight off-ecliptic movement.

Looking down on the ecliptic at some time t, from or from above the north pole of the ecliptic, we would see a triangle with the Sun at one vertex S, the Earth at another vertex E, and the superior planet at the third vertex P. We also need to consider a few other points in the plane of the ecliptic, notably the point O through which the planet passes when it is at opposition to Earth, i.e., when SEP is a straight line, not a triangle with finite area. We also need to consider the line FG shown in Fig. 6.21. This line is parallel to the line SO, goes through the center of the Earth, and it is convenient to choose the position of F such that the angle OSF is a right angle.

An observer on Earth looking along EG at time t will see the same star field as he or she would at the time t_{opp} of opposition looking along O, when the Earth

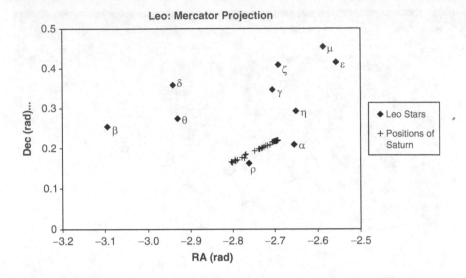

Fig. 6.13. Saturn moving within Leo. On this scale the movement cannot be seen in detail. The measurements started on 6 November 2007 and continued until 22 May 2008.

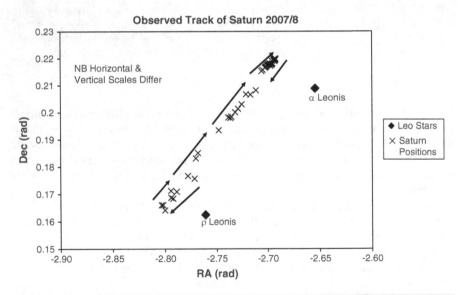

Fig. 6.14. Close-up of Saturn moving within Leo. The *arrows* indicate the directions of apparent motion of the planet. The change from eastward to westward motion and back is the so-called epicycle, which so confused the ancients.

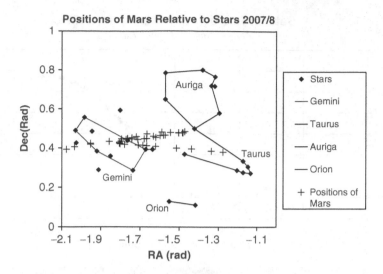

Fig. 6.15. Positions of Mars from 30 August 2007 to 2 May 2008. During this time, Mars appeared to move relative to the heavens much more than Saturn. It crossed from Taurus into Gemini, then went back almost to β Tauri (also known as γ Aurigae) before returning eastward through Gemini.

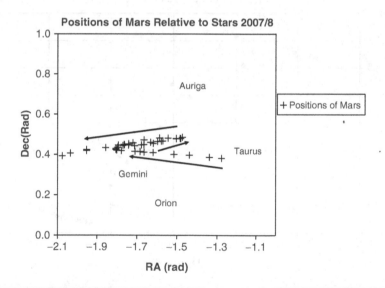

Fig. 6.16. Repeat of Fig. 6.15 with the stars removed for clarity. The *arrows* show the eastward (direct) motion and the intervening period of retrograde (westward) apparent motion. The retrograde motion was from 16 November 2007 to 30 January 2008. This was a shorter period of time than the retrograde motion of Saturn, which lasted 4½ months.

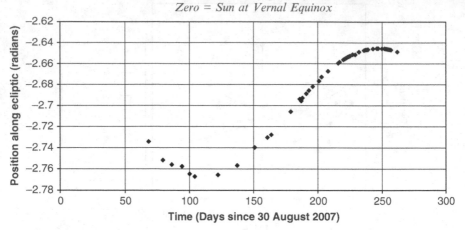

Fig. 6.17. 'Right ascensions' measured for Saturn along the ecliptic, not the celestial equator.

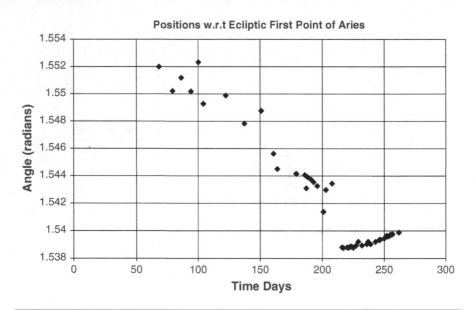

Fig. 6.18. Angles relative to ecliptic north pole measured for Saturn along the ecliptic, not the celestial equator. Notice the much greater precision obtained once I began using a telescope to track the planet after Day 215. If Saturn were on the ecliptic, the angle relative to the ecliptic north pole would be $\pi/2$ or 1.571 radians.

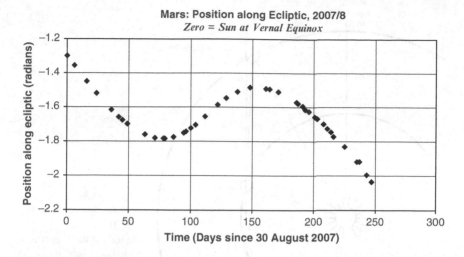

Fig. 6.19. "Right ascensions" measured for Mars along the ecliptic, not the celestial equator.

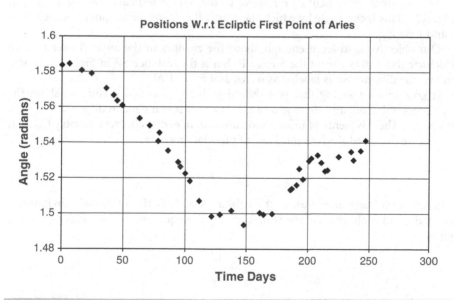

Fig. 6.20. Angles relative to ecliptic north pole measured for Mars along the ecliptic, not the celestial equator. If Mars were on the ecliptic, the angle relative to the ecliptic north pole would be $\pi/2$ or 1.571 radians.

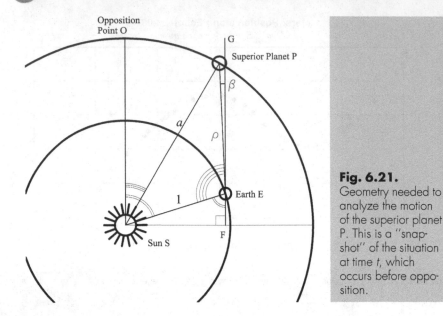

Fig. 6.21.
Geometry needed to analyze the motion of the superior planet P. This is a "snapshot" of the situation at time *t*, which occurs before opposition.

would be along the line SO. Indeed, the planet P would then appear to be in the direction SO.

The apparent movement of P relative to the stars is indicated by the angle β in Fig. 6.21. This is the angle by which the planet P will appear to move between time t and time t_{opp}.

Our objective is to learn enough about the changes in the angle β over time to calculate the distance from the Sun to P that is the distance SP in Fig. 6.21. In this figure, the distance SP is labeled as a AU, and SE as 1 AU.

It goes without saying that you should grab any knowledge you can about the dynamics of P from anywhere you can. It is not as if you have to do a black bag job on NASA. There is plenty of information around. In earlier chapters, notably Chap. 1, (1.19), we discussed Kepler's third law of planetary motion,

$$\frac{GM}{4\pi^2} = \frac{r^3}{T^2} = \text{constant}, \tag{6.11}$$

where r is the orbital radius, T the orbital period, G the universal gravitational constant, and M the mass of the Sun. Applying this equation to our current situation gives

$$\frac{a^3}{T_P^2} = \frac{1^3}{T_E^2}, \tag{6.12}$$

where the radii are in AU, so that

$$\frac{a^3}{1^3} = a^3 = \frac{T_P^2}{T_E^2} = \frac{2\pi/\omega_P^2}{2\pi/\omega_E^2} = \frac{\omega_E^2}{\omega_P^2}, \tag{6.13}$$

since the period T of an orbit is $2\pi/\omega$, where ω is the angular velocity of the planet's orbit. An immediate consequence of (6.13) is that

$$\omega_P a^{3/2} = \omega_E. \tag{6.14}$$

This in turn tells us what the angles OSP and OSE are in Fig. 6.21:

$$\angle OSE = \omega_E(t_{opp} - t) \quad \text{and}$$
$$\angle OSP = \omega_P(t_{opp} - t) = a^{2/3}\omega_E(t_{opp} - t) = a^{2/3}\angle OSE. \tag{6.15}$$

In Fig. 6.21, angle OSP is marked by a triple arc; and angle OSE is marked by a double arc, as is the angle SEF, which must be equal to angle OSE since angle ESF is equal to $((\pi/2)-$angle OSE); and since the angles of the right-angle triangle ESF must add up to π radians (see Appendix A).

The angle SEP is marked in Fig. 6.21 with a quadruple arc. Angles SEF, SEP, and PEG (labeled β) add up to π radians. We already know that $\angle SEF = \omega_E(t_{opp} - t)$, so

$$\angle SEF + \angle SEP + \angle PEG = \angle SEF + \angle SEP + \beta = \omega_E(t_{opp} - t) + \angle SEP + \beta = \pi.$$
$$\therefore \beta = \pi - \omega_E(t_{opp} - t) - \angle SEP. \tag{6.16}$$

If we can solve for angle SEP, we know β, the angle we are tying to calculate. Furthermore, given what we have said, we are quite likely to find a formula for β in terms of a, which is what we really want to know.

We also know ω_E. The earth's orbital period is 365.25 days, so

$$\omega_E = \frac{2\pi}{T_E} = \frac{2\pi}{365.25} = 0.01720 \text{ radians/day}. \tag{6.17}$$

Before proceeding, I will point out that from Fig. 6.21

$$\angle PSE = \angle OSE - \angle OSP = (\omega_E - \omega_P)(t_{opp} - t) = \omega_E(1 - a^{2/3})(t_{opp} - t). \tag{6.18}$$

The pieces are coming together. I am going to use two trigonometric tricks: the sine rule (http://en.wikipedia.org/wiki/Sine_rule) and the cosine rule (Appendix A), to nail down the angle SEP.

First, the cosine rule gives me the distance EP from the Earth to the superior planet, which for some reason seems to be called ρ (Greek lower case *rho*) in the celestial mechanics literature. This always makes me think that it is a bit too far to row a boat there. Anyway, I digress. The cosine rule, the cosine rule...applied to triangle PSE, it tells us that

$$\rho^2 = a^2 + 1^2 - (2 \times a \times 1 \times \cos(\angle PSE))$$

$$\therefore \rho = \sqrt{a^2 + 1 - (2a\cos(\angle PSE))} = \sqrt{a^2 + 1 - (2a\cos(\omega_E(1 - a^{2/3})(t_{opp} - t)))}, \tag{6.19}$$

where I have used (6.18) in the last line.

Right, now let us have a go with the sine rule on triangle PSE. As per the Appendix A, the conditions are satisfied for the obtuse angle version of the sine rule because the angle PES is obtuse; and lengths SE < SP and SF < SE. Therefore

$$\frac{a}{\sin(\pi - \angle\text{SEP})} = \frac{\sqrt{a^2 + 1 - (2a\cos(\omega_E(1 - a^{2/3})(t_{\text{opp}} - t)))}}{\sin(\omega_E(1 - a^{2/3})(t_{\text{opp}} - t))}. \qquad (6.20)$$

I can get angle SEP from this equation:

$$\arcsin\left(\frac{a\sin(\omega_E(1 - a^{2/3})(t_{\text{opp}} - t))}{\sqrt{a^2 + 1 - (2a\cos(\omega_E(1 - a^{2/3})(t_{\text{opp}} - t)))}}\right) = \pi - \angle\text{SEP}. \qquad (6.21)$$

We got there! Thank goodness for spreadsheets to do the grunt work of evaluating this equation. From (6.16),

$$\pi - \angle\text{SEP} - \omega_E(t_{\text{opp}} - t) = \beta. \qquad (6.22)$$

Substituting (6.21) into (6.22) gives

$$\beta = \arcsin\left(\frac{a\sin(\omega_E(1 - a^{2/3})(t_{\text{opp}} - t))}{\sqrt{a^2 + 1 - (2a\cos(\omega_E(1 - a^{2/3})(t_{\text{opp}} - t)))}}\right)$$
$$- \omega_E(t_{\text{opp}} - t). \qquad (6.23)$$

Finally, we have solved for β, the "right ascension" along the ecliptic relative to the ecliptical "right ascension" at opposition. Fortunately, (6.23) still works when $t > t_{\text{opp}}$.

We don't have a hope in you-know-where of turning this equation inside out to get a in terms of β. You just cannot do that when there are cosines and sines and stuff around. Besides, there is another unknown: the opposition time t_{opp}.

How do we get out of that, then?

There are no doubt many ways to skin this particular cat, but mine was to use a spreadsheet program (Excel) to try out a very large number of combinations of a and t_{opp}. In fact I tried out 32,768 combinations per planet. I then used the method of least squares to select the combination that best fitted my measured positions along the ecliptic. The method was exactly as in Chap. 4. It is described in Appendix B. The technique was to add all the values of $(\varphi_{\text{trial}} - \varphi_{\text{observed}})^2$, where φ_{trial} is the trial value of the "right ascension" ($\varphi = \varphi_{\text{opposition}} + \beta$) in the coordinate system of the ecliptic at the same time as the observation. The combination of a and t_{opp} that minimized $\sum(\varphi_{\text{trial}} - \varphi_{\text{observed}})^2$ was selected.

There are a three tricky bits to this. First, (6.23) is really too complicated to type straight into a spreadsheet cell. So wrote some Excel macros. I taught myself to do this from a book years ago. Since I do not do it very often, I have to refer a book every time.[37] I have never written more than a trivial macro in the other major spreadsheet, OpenOffice Calc, but the method is obviously very similar. If you truly hate macros, you could type (6.23) straight into spreadsheet cells, but it would not be much fun to debug this.

The second tricky bit is that if you wind the clock back enough, the angle SEP will cease to be acute. The same applies if you go far enough forward in time after the opposition. You then have to replace (6.21) with

$$\arcsin\left(\frac{a\sin(\omega_E(1-a^{2/3})(t_{opp}-t))}{\sqrt{a^2+1-(2a\cos(\omega_E(1-a^{2/3})(t_{opp}-t)))}}\right)=\angle SEP, \qquad (6.24)$$

so that (6.23) becomes

$$\beta=\pi-\arcsin\left(\frac{a\sin(\omega_E(1-a^{2/3})(t_{opp}-t))}{\sqrt{a^2+1-(2a\cos(\omega_E(1-a^{2/3})(t_{opp}-t)))}}\right) \qquad (6.25)$$
$$-\omega_E(t_{opp}-t).$$

The thought of solving for the transition point between the versions of β in (6.23) and (6.25) did not fill me with joy. However, I do know that the transition will occur when the arcsin in (6.23) and (6.25) goes through a maximum value of $\pi/2$. So I set my Excel macro to search for whether this arcsin was increasing or decreasing over very tiny time intervals, and use the IF-THEN-ELSE constructs to select the appropriate one.

The final bit of trickiness is that I need to add β to the position of the planet on the ecliptic at opposition. In Chap. 5, I worked out how to do this, and used the data I calculated there to give me a value of the "right ascension" angle along the ecliptic at every trial opposition time I examined.

So. . .how well did I do?

The actual time of the 2008 Saturn Opposition was 12:05:23 GMT on February 24th (http://telescopes.net/doc/3020). My time is 4½ h early, but over 200 days, I think that is pretty good accuracy.

Further evidence that I got the time of the opposition about right comes from Fig. 6.22, a composite of photos of Saturn taken before and after the opposition. The shadow cast by the planet onto its rings can be seen to move form the West side before the opposition to the East side afterward. This is consistent with the Earth overtaking Saturn while orbiting eastward.

From a Web site at http://homepage.ntlworld.com/mjpowell/Astro/Saturn-Orbit. htm, I estimate that Saturn at the 2008 opposition was about a quarter of the way

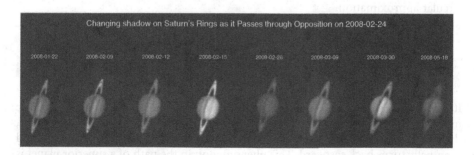

Fig. 6.22. The shadow cast by Saturn onto its rings moves from the West side to the East side as time progresses. You can see this shadow on the rings close to the edge of the planet as they go behind Saturn.

Table 6.1. Distances and Opposition Times Measured for Saturn

Planet	Distance (AU) from Sun	Opposition Time
Saturn	9.27	February 24, 2008 07:35 UT

Table 6.2. Distances and Opposition Times Measured for Mars

Planet	Distance (AU) from Sun	Opposition Time
Mars	1.68	December 25, 2007 06:46 UT

between perihelion and aphelion. From Wikipedia (http://en.wikipedia.org/wiki/Saturn) the perihelion is 9.05 AU from the Sun; and the aphelion is 10.12 AU away. A quarter of the way is a Sun-Saturn distance of 9.32 AU. Compared to this my value in Table 6.1 is about half a percent low. I am really pleased with that.

My opposition time for Mars (Table 6.2) was not quite so accurate as for Saturn, as expected. It was actually at 19:47 GMT on 24th December (http://seds.org/~spider/spider/Mars/marsopps.html), so I was 10 h 59 min late, out of about 200 days' observing.

There is an applet at http://www.windows.ucar.edu/tour/link=/mars/mars_orbit.html where you can watch the distance from Mars to Earth change. It turns out that my observations around the opposition of mars were not that far from its aphelion, whose distance from the Sun is 1.67 AU (http://en.wikipedia.org/wiki/Mars). My value is close to that. However, the orbit of Mars is much more elliptical than that of any planet except Mercury, so miracles cannot be expected. To put this into perspective, the perihelion is only 1.38 AU from the Sun. So, really, I have overestimated the Sun-Mars distance in Table 6.2. That is the price paid for the circular approximation.

Epicycles

It can be seen in Figs. 6.23 and 6.24 that the angle of the planet along the ecliptic moves at first toward the East (negative radians), reaching an easternmost point, then moves toward the West (radians getting less negative or more positive), and eventually turns back eastward. This phenomenon in the path of a superior planet is known as an epicycle.

I will now show that epicycles arise because both the Earth and the superior planet orbit the Sun; and because the Earth, orbiting the faster, overtakes the superior

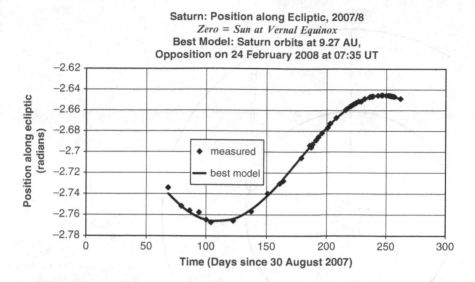

Fig. 6.23. Comparison of measured positions with the least-squares fitted circular orbit model.

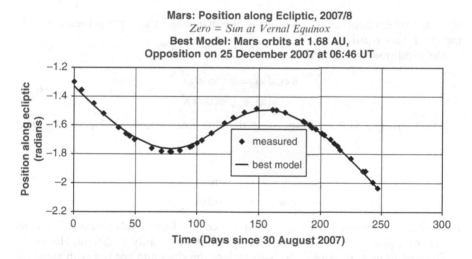

Fig. 6.24. Comparison of measured positions with the least-squares fitted circular orbit model.

planet once a year. An outline of this calculation is given in Tatum,[38] although he leaves the detailed workings, which I have included in full, to the reader (Fig. 6.25).

Tatum's way to calculate the positions of the stationary points in the planet's path is to note that the superior planet does not move relative to the background stars

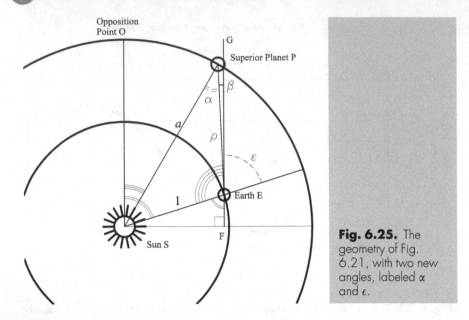

Fig. 6.25. The geometry of Fig. 6.21, with two new angles, labeled α and ϵ.

when the components of the motions of the planet and the earth perpendicular to the line EP are equal.

The condition for this is that

$$a\cos(\alpha) = a^{3/2}\cos(\varepsilon);$$
$$\cos(\alpha) = \sqrt{a}\cos(\varepsilon). \tag{6.26}$$

But from triangle SEP we have, using the sine rule (http://en.wikipedia.org/wiki/Sine_rule)

$$\frac{1}{\sin(\alpha)} = \frac{a}{\sin(\varepsilon)}; \tag{6.27}$$
$$a\sin(\alpha) = \sin(\varepsilon).$$

From these, I want to get rid of α (alpha), which is not very easily measured, and get a value of a (lower case A) in terms of ε, which is fairly easily measured. Here goes.

The first thing I propose to do is to replace the sines and cosines with tangents. Having followed the calculation, I know that it is worth doing this.

Remembering that

$$\sin\theta = \frac{\text{Opposite}}{\text{Hypotenuse}} = \frac{O}{H}, \quad \cos\theta = \frac{\text{Adjacent}}{\text{Hypotenuse}} = \frac{A}{H}, \text{ and}$$
$$\tan\theta = \frac{\text{Opposite}}{\text{Adjacent}} = \frac{O}{A}, \tag{6.28}$$

and that Pythagoras' theorem states that the square of the hypotenuse of a right-angled triangle is equal to the sum of the squares of the other two sides, i.e.,

$$O^2 + A^2 = H^2, \tag{6.29}$$

it follows from (6.28) that

$$\sin^2\theta + \cos^2\theta = \frac{O^2}{H^2} + \frac{A^2}{H^2} = \frac{O^2 + A^2}{H^2} = \frac{H^2}{H^2} = 1, \tag{6.30}$$

so that

$$\sin\theta = \frac{O/A}{H/A} = \frac{\tan\theta}{1/A\sqrt{A^2 + O^2}} = \frac{\tan\theta}{\sqrt{\frac{A^2}{A^2} + \frac{O^2}{A^2}}} = \frac{\tan\theta}{\sqrt{1 + \tan^2\theta}} \tag{6.31}$$

and

$$\cos\theta = \frac{A/A}{H/A} = \frac{1}{\frac{1}{A}\sqrt{O^2 + H^2}} = \frac{1}{\frac{1}{A}\sqrt{A^2 + O^2}} = \frac{1}{\sqrt{\frac{A^2}{A^2} + \frac{O^2}{A^2}}} = \frac{1}{\sqrt{1 + \tan^2\theta}}. \tag{6.32}$$

Hence, (6.26) becomes

$$\frac{1}{\sqrt{1 + \tan^2\alpha}} = \frac{\sqrt{a}}{\sqrt{1 + \tan^2\varepsilon}}, \tag{6.33}$$

and (6.27) becomes

$$\frac{a\tan\alpha}{\sqrt{1 + \tan^2\alpha}} = \frac{\tan\varepsilon}{\sqrt{1 + \tan^2\varepsilon}}. \tag{6.34}$$

Hence, using (6.33),

$$\frac{a\sqrt{a}\tan\alpha}{\sqrt{1 + \tan^2\varepsilon}} = \frac{\tan\varepsilon}{\sqrt{1 + \tan^2\varepsilon}}; \tag{6.35}$$
$$a\sqrt{a}\tan\alpha = \tan\varepsilon.$$

Equation (6.27) becomes, when combined with (6.26),

$$a\sqrt{1 - \cos^2\alpha} = \sin\varepsilon,$$
$$a\sqrt{1 - a\cos^2\varepsilon} = \sin\varepsilon, \tag{6.36}$$
$$\sqrt{1 - a\cos^2\varepsilon} = \frac{\sin\varepsilon}{a}.$$

Squaring both sides of (6.36) gives

$$1 - a\cos^2\varepsilon = \frac{\sin^2\varepsilon}{a^2}.$$

Now is the time to use the tangent formulae (6.31) and (6.32) to get rid of the sines and cosines.

$$1 - \frac{a}{1 + \tan^2\varepsilon} = \frac{\tan^2\varepsilon}{a^2(1 + \tan^2\varepsilon)},$$

$$1 + \tan^2\varepsilon - a = \frac{\tan^2\varepsilon}{a^2},$$

$$a^2 + a^2\tan^2\varepsilon - a^3 = \tan^2\varepsilon,$$

$$a^2 - a^3 = (1 - a^2)\tan^2\varepsilon, \qquad (6.37)$$

$$a^2(1 - a) = (1 - a^2)\tan^2\varepsilon,$$

$$a^2(1 - a) = (1 + a)(1 - a)\tan^2\varepsilon,$$

$$\frac{a^2}{(1 + a)} = \tan^2\varepsilon.$$

Finally, by taking the square root of (6.37), we end up with an equation for the tangent of ε, namely

$$\tan\varepsilon = \frac{a}{\sqrt{1 + a}}. \qquad (6.38)$$

Therefore

$$\varepsilon = \arctan\left(\frac{a}{\sqrt{1 + a}}\right), \qquad (6.39)$$

where ε is the angular distance between a stationary point and the opposition of the planet. The angular distance between the two stationary points will be

$$2\varepsilon = 2\arctan\left(\frac{a}{\sqrt{1 + a}}\right). \qquad (6.40)$$

How does (6.40) compare to what I observed? The easiest way to calculate this is to note that I already have an expression for $\pi-\varepsilon$, viz., (6.23). If I use (6.40), I calculate just under 136 days between the stationary points of Saturn, and just under 78 days between the stationary points of Mars.

I only observed one stationary point for Mars and one for Saturn. I estimated the dates for the stationary points I did not observe from Figs. 6.23 and 6.24. For Saturn, a time of 136 days between stationary points is consistent with Fig. 6.23; and with the widely published date of December 20, 2007 (http://www.poyntsource.com/New/Diary.htm) for the stationary point that I did not observe. For Saturn, I therefore measured a time between stationary points of about 134 days.

I observed Mars to be stationary on November 16, 2007, and the published value of the other stationary point is January 30, 2008 – a period of 75 days. This is not quite as good as the agreement between prediction and observation for Saturn, but it is not bad either.

Conclusion

Using my circular-orbit approximation, I measured a very good Sun-Saturn distance; and a reasonably good Sun-Mars distance. The reason for the less good result in the latter case is that Mars has a much less circular orbit than the other superior planets.

The retrograde motion phase of the epicycles seen in the motion of superior planets lasts approximately the time expected from analysis of a model of circular orbits.

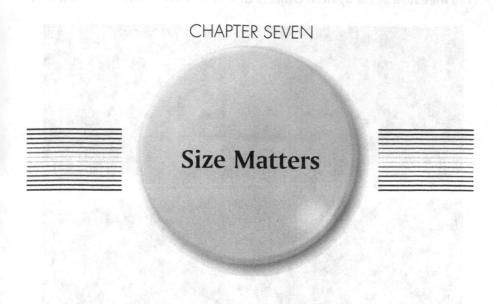

CHAPTER SEVEN

Size Matters

This is something I have not talked about much. There is a reason.

I needed to derive some geometrical and statistical results to get us this far. Our hardest work is now over. The thrust of this chapter is actually very simple. Since I know how big the planets look in my photos, and I now know how far away they are, I can calculate their sizes.

I also need the stuff in Chap. 4 about converting lengths on photos to angles in arcminutes, which I showed you how to do in (4.1)–(4.7). The only slight difference from this treatment is that when taking planet photos, I used a Barlow lens. You have to multiply the image size you calculate in arcminutes by the magnification of the Barlow. Do remember to calibrate your Barlow: the quoted magnifications are not exact.

Do not forget that the Planet–Sun distance is not the Planet–Earth distance. The latter varies all the time as both planets orbit. This is obvious from, say, Fig. 1.12. The effect is less obvious in Fig. 6.22, but it is there. The rightmost image of Saturn is about 8% smaller than the one taken nearest Opposition, the fourth from the right. In the case of Mars, Fig. 7.1 shows that the Earth-Mars distance varies a lot. (My wife pointed out that it could be Mars that changes size, but not even she believes that.)

We cannot escape the need to allow for the varying distance to these planets if we want to measure and calculate their sizes.

For superior planets, the key diagram is Fig. 6.21. The required distance is labeled as ρ in this figure. I calculated its value to be that in (6.19). For Mars and Saturn, I know the orbital radius a and opposition time t_{opp}. They are given in Tables 6.1 and 6.2. I can simply plug these values into (6.19) and calculate the distance to the planet when the photograph was taken.

For the inferior planets Mercury and Venus, the principle is very similar. We now use Fig. 7.2 to define the distance ρ to the inferior planet. I also make minor alterations to (6.19) to produce an equation for the distance ρ to the inferior planet:

J.D. Clark, *Measure Solar System Objects and Their Movements for Yourself!*,
Patrick Moore's Practical Astronomy Series,
DOI: 10.1007/978-0-387-89561-1_7, © Springer Science + Business Media, LLC 2009

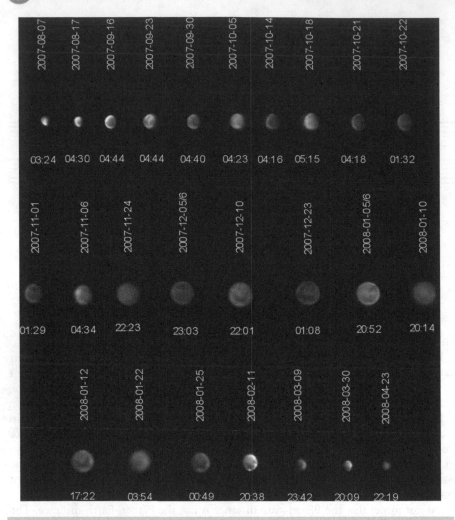

Fig. 7.1. Photographs of Mars taken during the 2007/8 apparition. Notice how much the apparent size of the planet varies.

$$\rho = \sqrt{1 + a^2 - (2a\cos(\omega_E(a^{2/3} - 1)(t_{IC} - t)))}, \qquad (7.1)$$

where now $a < 1$ so that $(a^{2/3} - 1)$ is positive. The time t_{IC} is the time of the inferior conjunction. As with (6.19), I have applied the cosine rule, (A.44) in the Appendix A, to Fig. 7.2.

For both superior and inferior planets, we need to turn our knowledge of ρ and the angle θ subtended by the planet on our photos into a size for the planet.

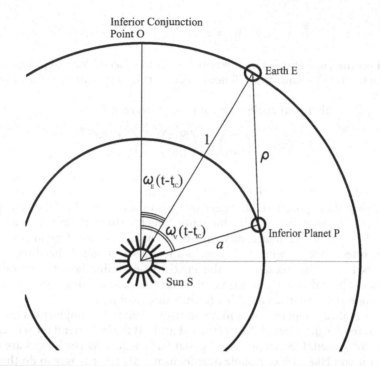

Fig. 7.2. Geometry needed to work out the distance to the inferior planet P. This is a "snapshot" of the situation at time *t*, which occurs before inferior conjunction (IC).

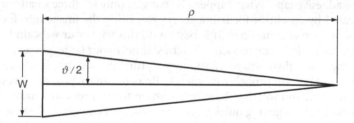

Fig. 7.3. The parameters necessary to turn knowledge of the angle θ and the distance ρ into a size W for our planet.

Figure 7.3 shows you how to do this. The geometry follows directly from the definition of a tangent in Appendix A, (A.1):

$$\frac{(W/2)}{\rho} = \tan\left(\frac{\theta}{2}\right) \tag{7.2}$$

In the pre-spreadsheet era, you would have solved (7.2) for *W* by making the small angle approximation in (A.5). If you are not comfortable doing this, you can just let your favorite spreadsheet solve the equation

$$W = 2\rho \tan\left(\frac{\theta}{2}\right) \tag{7.3}$$

You will get the same answer either way to a large number of decimal places so long as the angle θ is in radians. You will need to convert it to radians in a spreadsheet in any case.

Do not forget that your angle θ may not be in degrees:

$$1 \text{ radian} = \left(\frac{\pi}{180}\right)^\circ = \left(\frac{60\pi}{180}\right)' = \left(\frac{3,600\pi}{180}\right)'' \tag{7.4}$$

To measure my angles θ in the photos in Figs. 1.12, 6.22, and 7.1, I used CAD software. I drew circles around the visible parts of the planets using the "draw a circle by specifying three points" tool. There is a bit of an art to this, and it requires practice, as I discovered when doing the same thing for Moon pictures. I always drew three circles around each image, each on a different layer so that I could not see the previous ones while drawing each one. Sometimes I averaged the three results, sometimes I took the middle one (the *median*) depending how the mood took me. It makes little difference. Just do not expect the variance, or its square root the standard deviation, to mean anything for just three samples.

There was always a lot of scatter in my measurements. My equipment is not good enough to resolve the edges of these planets sharply. It is a lot better if the telescope is properly collimated. I recommend doing that fairly regularly. The images are also a lot better if you take a bit of trouble over focusing. My favorite way to do this is to use a baffle with three holes where the main front cap goes on my telescope. You can buy these things for an obscene price for a bit of plastic you might step on and break in the dark, or you can do what I did, which is to make one out of cardboard and waterproof adhesive tape. What happens is that you only see three small parts of the "blob" created by an out-of-focus image. As you bring the image into focus these three small blobs merge into one. It is best to do this with your webcam brightness turned right up initially, and to focus a bright star near your target planet rather than a planet. Once your three images have merged, turn the webcam brightness down, and the three blobs will probably de-merge. Refocus, and repeat until you can no longer see the star. You will than have pretty sharp focus. I reckon to take 20 min to do this, although I am getting quicker with practice. An engineer colleague of mine, Blair McKnight, once took my rack-and-pinion focuser apart and cured the backlash in it by wrapping plumber's Teflon tape around the barrel of the focuser. Life got a lot better after that. Finally, if your three images would not merge, you probably need to recollimate your telescope.

Averaging the diameter measurements over a lot of photos does seem work quite well as a way of dealing with fuzzy pictures.

How often do you need to recollimate? That depends how much you move the scope. It is carrying it around that does the misaligning; and it is usually the odd bump because you drive over a pothole rather than regular, careful transportation. My worst experience that way was when I put my 8-in. Newtonian onto the mount, and as I walked away, it fell to the ground with a sickening clang as it bounced off the tripod leg. It rolled around just enough that it landed focuser first onto the grass, and stove the tube in very nicely. I tried hammering it back into shape, but one nut holding the secondary mirror's spider onto the tube would not undo, so I did not dare hammer hard. In the end, I worked out a technique for placing G-clamps of various sizes through the focuser hole, and pulling the tube aluminum straight. There is quite a scar on the tube, but it recollimates just fine.

I forgot to mention that I did not measure the date of Venus' 2007 inferior conjunction. That in turn means that I did not know ρ, the instantaneous Earth–Venus distance. I dealt with this problem in a simple but stupid way. If I *assumed* a conjunction date, I could calculate the diameters for each Venus photo. They should be constant, apart from random scatter due to the variable quality of my photography. Therefore, if I choose the correct conjunction date, there would be no systematic error due to a wrong conjunction date to add to the random errors (see Appendix B). So the calculated standard deviation should be a minimum, since it really sums random and systematic errors. Therefore I took the measured conjunction date to be the one that gave me the minimum standard deviation in my Venus diameters. It was August 19, 2007 at 13:08, cf. the Cambridge Planetary Handbook's August 18, 2007.[39] I used a Monte Carlo technique to determine this (see Appendix B) in which I let Microsoft Excel work out the standard deviation of all the diameters for several thousand assumed times of inferior conjunction. Two lots of 8,192 instances gave the same time to five significant figures.

Anyway, back to my distance measurements. The values I obtained are in Table 7.1. They are given in astronomical units. Even though this is unusual, I make no apology for it. You can see how tiny the planets are compared to the distances between them. The planetary diameters are of the order of one ten-thousandth of an astronomical unit. The planets themselves go out to about 30 AU, so the diameter of the planetary Solar System is of the order of 60 AU. That is about a million times the diameter of Venus.

Given that the current definition of a planet requires it to absorb most of the interplanetary matter in its neighborhood, the above reasoning shows that they are awfully small vacuum cleaners.

Oops, I forgot to mention that the disk of Saturn is not very circular. I measured it by the following cunning means. In the CAD system, I drew a straight line freehand between the tips of the rings. I then drew an ellipse, using the snap tools to force its center to be the midpoint of the line between the rings. I made its major axis be along this line, and placed the ends of the major axis at the edge of the planet proper. The minor axis I made go from one pole to the other. Since the planet is tilted relative to the Earth, this is only approximately true. When in late 2009 the rings are edge-on, I intend to repeat this measurement. A good value of the ratio of polar to equatorial diameters will in principle enable me to estimate the rotation speed, since the surface is where centrifugal and gravitational forces balance. The estimate will be crude, because the rotation speed is not uniform, but I cannot think of any other way to reveal this rotation speed with typical amateur equipment.

I have not of course managed to measure an astronomical unit in miles, but this is an easy number to look up.[40] Using the values

$$1 \text{ AU} = 1.496 \times 10^8 \text{ km} = 9.296 \times 10^7 \text{ miles}, \qquad (7.5)$$

I have checked the values in Chap. 2, (2.1) and in Table 7.1 against those in the Cambridge Planetary Handbook,[41] and report the comparisons in Table 7.2.

Table 7.1. Planetary Diameters in AU

Planet	Venus	Mars	Saturn
Diameter (AU) (±standard deviation)	$7.38 \pm 0.54 \times 10^{-5}$	$4.83 \pm 0.41 \times 10^{-5}$	Equatorial: $7.98 \pm 0.61 \times 10^{-4}$ Polar: $6.87 \pm 0.32 \times 10^{-4}$ Rings: $1.74 \pm 0.08 \times 10^{-3}$

Table 7.2. Planetary Diameters in AU

Planet	Diameter (AU) (±Standard Deviation)	Values in Reference 41 (AU)	Error of Mean
Venus	$7.38 \pm 0.54 \times 10^{-5}$	8.09×10^{-5}	−3.3%
Earth	$8.33 \pm 0.76 \times 10^{-5}$	8.51×10^{-5}	−2.2%
Mars	$4.83 \pm 0.41 \times 10^{-5}$	4.53×10^{-5}	6.3%
Saturn	Equatorial: $7.98 \pm 0.61 \times 10^{-4}$ Polar: $6.87 \pm 0.32 \times 10^{-4}$ Rings: $1.74 \pm 0.08 \times 10^{-3}$	Equatorial: 8.05×10^{-4} Polar: 7.19×10^{-4} Rings: 1.73×10^{-3}	Equatorial: −1.0% Polar: −4.7% Rings: 0.7%

The biggest error is in the estimate of Mars' diameter. This reflects the greater uncertainty in its distance. The next biggest discrepancy is in Saturn's polar diameter. I have already mentioned the reason for the uncertainty here. All the published values lie within my error bars.

So the main part of my story ends on a high note: with amateur kit like mine, you can measure good values of planetary sizes.

Appendix A: Geometrical Appendix: Geometry for Those Who Have Forgotten

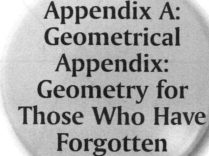

Basic Trigonometry

My working assumption is that most readers have a vague idea that trigonometry is about triangles and cosines and stuff, but cannot remember the details.

The purpose of this section is to remind such readers of some things they may have forgotten.

The first result I want to prove is that the sum of angles of a triangle is 180° or π radians. In Fig. A.1, angle A = angle F, angle B = angle D, and angle C = angle E. Since angles D, E, and F obviously add up to 180° or π radians, the same must be true of angles A, B, and C. I use this result a lot.

The definitions of sine, cosine, and tangent are as follows.

$$\sin(A) = \frac{\text{opposite}}{\text{hypotenuse}}$$

$$\cos(A) = \frac{\text{adjacent}}{\text{hypotenuse}}$$ (A.1)

$$\tan(A) = \frac{\text{opposite}}{\text{adjacent}} = \frac{\text{opposite/hypotenuse}}{\text{adjacent/hypotenuse}} = \frac{\sin(A)}{\cos(A)}.$$

The last line of (A.1) follows since the hypotenuse cancels out of the division.

Remember that the angles of a triangle add up to 180°. From this it follows that the angles in the triangle in Fig. A.2 are A, 90° and $(90° - A)$. It follows from (A.1) that

$$\sin(A) = \cos(90° - A) \quad \text{and} \quad \cos(A) = \sin(90° - A)$$ (A.2)

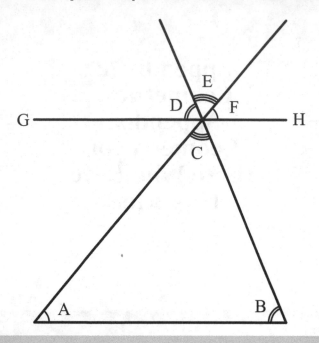

Fig. A.1. The angles of a *triangle* sum to 180° or π radians.

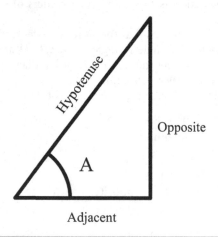

Fig. A.2. A *right angle triangle* showing the names of the sides relative to the angle A.

By the way, since you cannot divide by zero, you have a problem with (A.1) if cos (A) = 0. It turns out that this is true when A = 90°. Tan(90°) is normally reported in tables as being equal to infinity. It would be more accurate to state that as (A) approaches 90°, *tan(A)* gets bigger and bigger. Table A.1 shows what I mean.

Table A.1. Showing How Tan(A) Tends to Infinity as A Increases to 90°

A	Tan(A)
80°	5.67
89°	57.29
89.9°	572.96
89.99°	5,729.58
89.999°	57,295.78
89.9999°	572,957.80

What Happens if A is bigger than 90°?

Think of a circle whose radius is equal to the hypotenuse, just like the one shown in Fig. A.3. Now imagine that the x- and y-coordinates are as shown in Fig. A.3. The "Adjacent" side is in the x-direction; the "Opposite" side is in the y-direction.

If the angle A is between 90° and 180°, the usual practice is to take a bit of a liberty with what we mean by the "Adjacent" side. Imagine the geometry shown in Fig. A.4, and simply use the definitions of sine, cosine, and tangent we had before. It is also customary to note that the "Adjacent" side is along the negative part of the x-axis and therefore to give the length of the "Adjacent" side a negative value. Since all we are doing is defining things, we can define them how we darn well like, even if this does result in a rather odd use of the word "Adjacent."

Our right angle triangle is taken into the regions where A lies between 180° and 270°; and 270° and 360° in the same way as before. This is shown in Fig. A.5.

Between 180° and 270°, both the "Adjacent" and "Opposite" sides are given negative length. Between 270° and 360°, the "Adjacent" side is given positive length, and the "Opposite" side is given negative length.

Figures A.5 and A.6 show what $\sin(A)$, $\cos(A)$, and $\tan(A)$ look like if we use the above definitions.

In principle, these definitions can be extended to any angle, simply by taking the angle A round and round the circle of radius equal to the hypotenuse. Negative angles are no problem either: just take the angle A going clockwise instead of anticlockwise (Fig. A.7).

Angular Units

Finally, units are never as simple as they might be. Degrees came originally from ancient Babylon, where they used base 60 in their mathematics. There are decimal based angular units, but I never see them used. The unit which is commonly used in

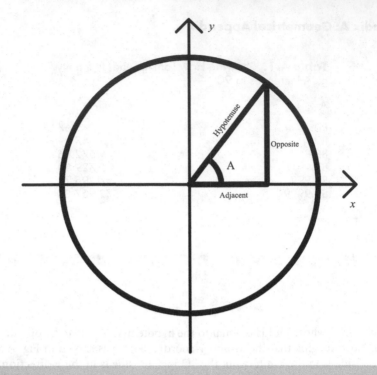

Fig. A.3. A *right angle triangle* showing the names of the sides relative to the angle A, a *circle* whose radius is equal to the hypotenuse, and x- and y-coordinates. The "Adjacent" side is in the x-direction; the "Opposite" side is in the y-direction.

mathematics is called the radian. It is defined so that there are 2π radians in a complete circle:

$$2\pi \text{ radians} = 360°$$
$$\pi \text{ radians} = 180°$$
$$\frac{\pi}{2} \text{ radians} = 90° \qquad \text{(A.3)}$$
$$1 \text{ radian} = \left(\frac{180}{\pi}\right)°$$

You might well ask why anyone would want to do this. The answer is that a lot of mathematics becomes a lot simpler if you use these units. They are a kind of "natural" unit for trigonometry and calculus.

One very useful approximation that only works in radians is that for angles below about 10°, i.e., below about 0.2 radians,

$$\sin(A) \approx A;$$
$$\cos(A) \approx 1; \qquad \text{(A.4)}$$
$$\tan(A) \approx A.$$

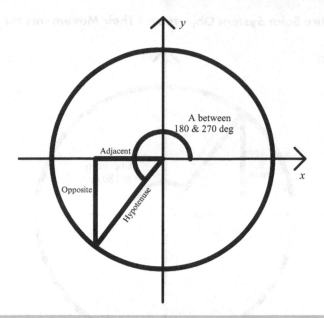

Fig. A.4. If the angle A is between 90 and 180°, the usual practice is to take a bit of a liberty with what we mean by the "Adjacent" side, imagine the geometry shown, and simply use the definitions of sine, cosine, and tangent we had before. It is also customary to note that the "Adjacent" side is along the negative part of the x-axis and therefore to give the length of the "Adjacent" side a negative value. Since all we are doing is defining things, we can define them how what we jolly well like, even if this does result in a rather odd use of the word "Adjacent."

Finally, there are two notations which mean "the angle whose sine (or cosine or tangent) is A". They are

$$\arcsin(A) = \sin^{-1}(A) = \text{angle whose sin is } A;$$
$$\arccos(A) = \cos^{-1}(A) = \text{angle whose cos is } A; \qquad (A.5)$$
$$\arctan(A) = \tan^{-1}(A) = \text{angle whose tan is } A.$$

Coordinate Systems

There are two coordinate systems of interest to astronomers: Cartesian and polar.

First, let us look at two-dimensional coordinates. Figure A.8 shows the coordinates of the point P in Cartesian or (x, y) coordinates. It also shows the coordinates of P in polar coordinates, (r, φ), where φ is the Greek lower case letter *phi*. The distance from the origin O to P is r. The angle the line OP makes with the x-axis is φ. By convention φ is positive going from $+x$ toward $+y$. Equally conventionally but equally arbitrarily, the $+x$ direction is toward the right and the $+y$ direction is upward.

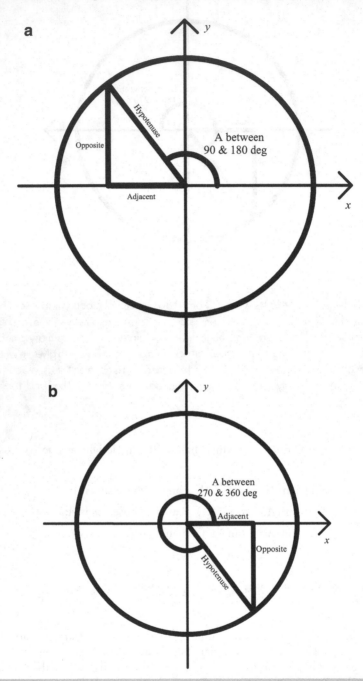

Fig. A.5. Our right angle triangle is taken into the regions where A lies between 180° and 270°; and 270° and 360° in the same way as before.

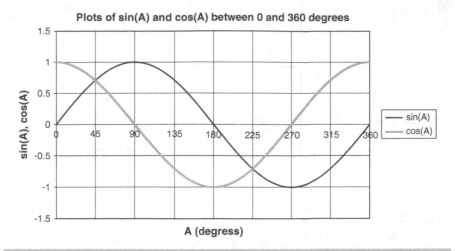

Fig. A.6. Plots of sin(A) and cos(A) for A between 0° and 360°.

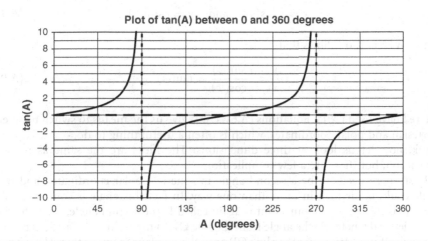

Fig. A.7. Plots of tan(A) for A between 0° and 360°. Note that tan(45°) = tan(225°) = +1, tan(135°) = tan(315°) = −1 and tan(0°) = tan(180°) = tan(360°) = 0.

It follows from Pythagoras' theorem that

$$r^2 = x^2 + y^2 \tag{A.6}$$

It also follows from (A.1) that

$$\frac{x}{r} = \cos(\varphi) \quad \text{and} \quad \frac{y}{r} = \sin(\varphi) \tag{A.7}$$

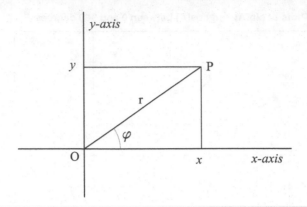

Fig. A.8. Cartesian and polar coordinates in two dimensions. The point P has Cartesian coordinates (x, y) and polar coordinates (r, φ).

Therefore

$$x = r\cos(\varphi) \quad \text{and} \quad y = r\sin(\varphi) \tag{A.8}$$

From (A.8) it also follows that

$$\frac{y}{x} = \frac{r\sin(\varphi)}{r\cos(\varphi)} = \tan(\varphi) \tag{A.9}$$

where I have used (A.1). Equations (A.6)–(A.9) give us the means to switch between Cartesian and polar coordinates, which is often a useful thing to do.

Ok, now let us look at three dimensions. The ideas are the same as in two dimension, but there are a few complications.

First, there needs to be a second angle in spherical polar coordinates, and this angle could be defined in more than one way. In fact this angle is conventionally defined as in Fig. A.9. If our point is again called P, the second angle, called by the Greek letter θ (theta), is the angle from the line OP to the z-axis. The angle φ is now the angle of the *projection* of the line OP onto the xy-plane from +x toward +y. Thus φ and θ are a kind of longitude and latitude. The latitude is not quite θ, but is $(90° - \theta)$ or $((1/2)\pi - \theta)$ depending whether you prefer degrees or radians. Thus, on the celestial sphere (see below), the declination would be $(90° - \theta)$ or $((1/2)\pi - \theta)$.

Ok. We are getting there. There is one other little joker in the three-dimensional coordinate deck: handedness.

If you imagine making your thumb, index finger, and second finger all straight and mutually perpendicular, you can then make x point along your index finger, y point along your second finger, and z point along your thumb (Fig. A.10). Notice that, if you do this with both hands, you cannot superimpose the two sets of fingers. If you align x and y, the z s will point in opposite directions, and so on. Left-handed and right-handed coordinate systems are different.

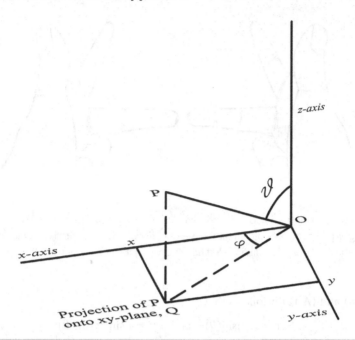

Fig. A.9. Cartesian and polar coordinates in three dimensions. The point P has Cartesian coordinates (x, y, z) and polar coordinates (r, θ, φ), where θ is lower case Greek *theta*.

You will have to put the book down, but if you have never done this before, it is a worthwhile exercise. Although it is normal to choose a right-handed coordinate system, I found that it was easier to analyze the celestial sphere in a left-handed one when I was working out the distances to superior planets, so that is what I did.

Returning to Fig. A.9, we can see that from Pythagoras' theorem

$$OQ^2 = x^2 + y^2 \qquad (A.10)$$

where Q is the Projection of P onto the *xy*-plane. Hence, applying Pythagoras' theorem a second time

$$OP^2 = OQ^2 + QP^2$$
$$\text{i.e., } OP^2 = x^2 + y^2 + z^2 \qquad (A.11)$$
$$\text{i.e., } r^2 = x^2 + y^2 + z^2$$

In three-dimensional coordinates, the length of line OP is r. The projection of OP onto the *xy*-plane is obtained by noting that

$$\frac{OQ}{OP} = \cos\left(\frac{\pi}{2} - \theta\right) = \sin(\theta) \qquad (A.12)$$
$$\text{i.e., } OQ = r\sin(\theta)$$

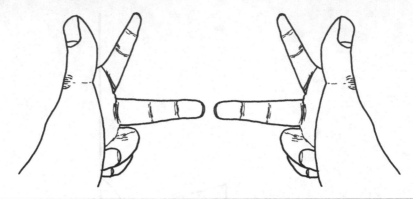

Fig. A.10. Showing how to make the fingers of your hand like coordinate axes. The *index finger* points in the *x*-direction, the *second finger* along *y* and the *thumb* along *z*. Left-handed coordinates are like the left hand; and right-handed ones like the right hand.

From (A.8) and (A.12) it follows that

$$x = r\cos(\varphi)\sin(\theta) \quad \text{and} \quad y = r\sin(\varphi)\sin(\theta) \tag{A.13}$$

Since

$$\frac{QP}{OP} = \cos(\theta) = \frac{z}{r}, \tag{A.14}$$

it follows that

$$z = r\cos(\theta). \tag{A.15}$$

Equations (5.7) and (A.15) enable us to extract Cartesian coordinates from spherical polar coordinates.

How do we go the other way? We already know from (A.11) how to work out r given x, y, and z. We get φ from (A.16):

$$\frac{y}{x} = \tan(\varphi), \quad \text{so } \varphi = \arctan\left(\frac{y}{x}\right) \tag{A.16}$$

We get θ from (A.11) and (A.15):

$$z = \sqrt{x^2 + y^2 + z^2}\sin(\theta), \quad \text{so } \theta = \arcsin\left(\frac{z}{\sqrt{x^2 + y^2 + z^2}}\right) \tag{A.17}$$

We now have formulae to transform both ways between three-dimensional Cartesian and spherical polar coordinates.

Proof of Coordinate Transform Formulae

These formulae are widely asserted in mathematics and other textbooks, but rarely derived or proved. I recently had to derive them to help my teenage daughter with her homework. I was taken aback by how long it took me to do this – a good two hours. So I felt it only fair to save you, dear reader, the trouble to which I had to go. I hope that understanding my derivation will take you considerably less than two hours.

My wife, whose double major was in mathematics and physics, and I agreed afterward that neither of us had been shown convincing derivations of these formulae at college (Fig. A.11).

The question is: what are the (x', y') coordinates of the point (x, y)?

The length OA is x, the length OF is y, the length OC is x', and the lengths OG and CE are both y'.

By simple trigonometry (see above)

$$OB = OC \cos(\psi) \qquad (A.18)$$

and

$$AB = DC = EC \sin(\psi) \qquad (A.19)$$

But

$$OA = OB - AB, \quad \text{or}$$
$$x = OC \cos(\psi) - EC \sin(\psi), \quad \text{or} \qquad (A.20)$$
$$x = x' \cos(\psi) - y' \sin(\psi).$$

Also by simple trigonometry

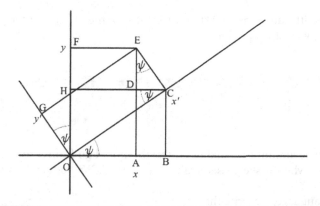

Fig. A.11. Showing two coordinate axes separated by an angle ψ (Greek lower case "psi"). The x-axis is shown horizontal and the y-axis is shown vertical. The x'-axis goes toward the top right and the y'-axis toward the top left. English letters are used to denote points in the diagram.

$$OH = OC\sin(\psi), \tag{A.21}$$

and

$$HF = DE = EC\cos(\psi). \tag{A.22}$$

But

$$OF = OH + HF, \text{ or}$$
$$y = OC\sin(\psi) + EC\cos(\psi), \text{ or} \tag{A.23}$$
$$y = x'\sin(\psi) + y'\cos(\psi).$$

We are sort of there, but I have it the wrong way round: I know (x', y') in terms of (x, y), when I really want them the other way round. That actually is not a big deal. From (A.20) and (A.23), I can write that

$$x\cos(\psi) + y\sin(\psi) = x'\cos^2(\psi) - y'\sin(\psi)\cos(\psi) + x'\sin^2(\psi)$$
$$+ y'\cos(\psi)\sin(\psi)$$
$$= x'\cos^2(\psi) + x'\sin^2(\psi) = x'. \tag{A.24}$$

I have simply multiplied my formula for x by $\cos(\psi)$, my formula for y by $\sin(\psi)$, added them, cancelled out some terms and used the well-known formula that

$$\cos^2(\psi) + \sin^2(\psi) = 1. \tag{A.25}$$

Now let me multiply my formula for x by $\sin(\psi)$, my formula for y by $\cos(\psi)$, and this time subtract them. I get

$$y\cos(\psi) - x\sin(\psi) = y'\cos^2(\psi) + x'\sin(\psi)\cos(\psi) - x'\cos(\psi)\sin(\psi)$$
$$+ y'\sin^2(\psi)$$
$$= y'\cos^2(\psi) + y'\sin^2(\psi) = y'. \tag{A.26}$$

I have done it! Equations (A.24) and (A.26) give me (x', y') in terms of (x, y), which is exactly what I was after.

Vectors: A Way of Simplifying Geometry

Imagine my wife and myself walking in straight lines. I go from A to B in Fig. A.12, and then to C, whereas she goes straight from A to C. We both end up in the same place.

In vector language, we write this as

$$\overrightarrow{AB} + \overrightarrow{BC} = \overrightarrow{AC}. \tag{A.27}$$

Now, suppose she walks from C to A and then to B, whereas I retrace my steps to B. There is a vector equation for this too:

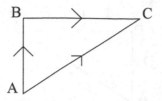

Fig. A.12. If I walk from A to B, and then walk from B to C (both in *straight lines*), whereas my wife walks in a straight line from A to C, we both end up in the same place.

$$\overrightarrow{CB} = \overrightarrow{AB} - \overrightarrow{AC}. \qquad (A.28)$$

AC is now negative, because she went from C to A. Rocket science it ain't.

We have taken account here of both the distances and the directions of the lines AB, BC, and AC. In (A.27) and (A.28), $\overrightarrow{AB}$, $\overrightarrow{BC}$, and $\overrightarrow{AC}$ are referred to as *vectors*. They have magnitude (the distance) and direction.

Unfortunately, there are many vector notations.

Alternative ways used to write (A.27) include:

$$\begin{aligned} \overrightarrow{AB} + \overrightarrow{BC} &= \overrightarrow{AC} \\ \underline{D} + \underline{E} &= \underline{F} \\ \underset{\sim}{D} + \underset{\sim}{E} &= \underset{\sim}{F} \\ \mathbf{D} + \mathbf{E} &= \mathbf{F} \end{aligned} \qquad (A.29)$$

The single letter is sometimes underlined, and sometimes bold. You cannot easily handwrite bold font, so this notation is usually restricted to printing. When I handwrite vectors, I usually use a curly underline, because this is harder to mistake for a slip of the pen. A straight underline is obviously more word-processor friendly.

Ok, my wife and I have done adding and subtracting. What happens when we multiply? Perhaps not. No jokes about my needing size and direction, please. We had better stick to multiplying vectors. There are two ways of doing this. One yields a vector, and the other yields a common or garden number. But you have much more street cred in the vector world if you call it a scalar instead of a number.

In this book, I have used the one that gives me a number. It is variously called the scalar product and the dot product of two vectors. That is because this product is often written as

$$\underline{D} \bullet \underline{E} = s, \qquad (A.30)$$

where s is a scalar quantity, a.k.a. a number. How do we actually do the multiplication? This will look very arbitrary, but I promise you it is a useful thing to do.

$$\underline{D} \bullet \underline{F} = |\underline{D}||\underline{F}|\cos(\psi) = DF\cos(\psi). \qquad (A.31)$$

the notation $|\underline{D}|$ means the length of $\underline{D}$. It is sometimes written D, without the underline, bold, or whatever. But we are not going to call it the length. Oh no. That would be much too easy. We are out to confuse you here. So we are going to call it the

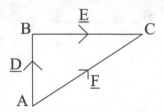

Fig. A.13. Showing the included angle ψ between the vectors $\underline{D}$ and $\underline{F}$.

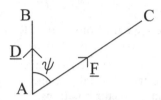

Fig. A.14. The normal convention is to label the x-component of the vector $\underline{W}$ as W_x and the y-component as W_y, with no *underlining* or *bolding*.

modulus of $\underline{D}$. There is a sort of legitimacy here. After all, people write the modulus of minus three as $|-3| = 3 = |+3|$. The term and the notation convey the idea of size or magnitude. Note that in (A.31), the order of $\underline{D}$ and $\underline{F}$ does not matter:

$$\underline{D} \bullet \underline{F} = \underline{F} \bullet \underline{D}. \qquad (A.32)$$

I chose to show you the definition for $\underline{D}$ and $\underline{F}$, not $\underline{D}$ and $\underline{E}$ because $\underline{D}$ and $\underline{E}$ are perpendicular, and the cosine of a right angle is zero. Hence

$$\underline{D} \bullet \underline{E} = 0. \qquad (A.33)$$

I use this property to calculate the position of the Sun at sunrise and sunset in Chap. 5, because the Sun, being exactly on the horizon, is then perpendicular to the zenith.

Note also that in Fig. A.13, from (A.1)

$$\frac{AB}{AC} = \cos(\psi). \qquad (A.34)$$

We can exploit this fact to set up some components in Cartesian coordinates for vectors. Figure A.14 shows how, at least in two dimensions.

It can be seen from Fig. A.14 that

$$W_x = |\underline{W}|\cos(\psi), \quad \text{and} \quad W_y = |\underline{W}|\sin(\psi). \qquad (A.35)$$

By the way, I can now come clean about the other requirement for a quantity to be a vector. It must not only have magnitude and direction, but also obey the coordinate transformation rule derived in the previous section. This may sound like a dry technicality, but if it did not hold, believe me, there would be chaos in astronomy.

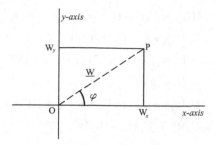

Fig. A.15. Extension of the Cartesian notation for vectors to three dimension. The angles θ, α, and β are the angles between the vector $\underline{W}$ and the axes. Since θ is one of the angles used to define spherical polar coordinates, I thought it is silly to rename it gamma.

The extension of the coordinate notation for vectors to three dimensions is straightforward (Fig. A.15).

I have one other very useful type of vector to define: a so-called *unit vector*. Unit vectors all have length one, but can point in any direction. For example, the unit vector pointing in the direction of $\underline{W}$ is

$$\hat{\underline{W}} = \frac{\underline{W}}{|\underline{W}|}. \tag{A.36}$$

In other words I have divided the vector $\underline{W}$ by the scalar $|\underline{W}|$ to give me a vector whose length is unity. Unit vectors are usually denoted by having a "∧" symbol on top, and are usually pronounced "double-you-hat," etc. Three especially useful unit vectors are $\hat{\underline{x}}$, $\hat{\underline{y}}$, and $\hat{\underline{z}}$. They point along our Cartesian axes. Since they are mutually perpendicular, it follows from (A.31) that

$$\hat{\underline{x}} \bullet \hat{\underline{y}} = \hat{\underline{x}} \bullet \hat{\underline{z}} = \hat{\underline{y}} \bullet \hat{\underline{z}} = 0. \tag{A.37}$$

and

$$\hat{\underline{x}} \bullet \hat{\underline{x}} = \hat{\underline{y}} \bullet \hat{\underline{y}} = \hat{\underline{z}} \bullet \hat{\underline{z}} = 1 \times 1 \times \cos(0) = 1. \tag{A.38}$$

I can, if the mood takes me, write $\underline{W}$ as

$$\underline{W} = W_x\hat{\underline{x}} + W_y\hat{\underline{y}} + W_z\hat{\underline{z}}. \tag{A.39}$$

This is the usual form in which vectors are written in Cartesian coordinates. It is generally taken that a vector with the same magnitude and direction as $\underline{W}$, but which does not start at the coordinate origin, is equal to $\underline{W}$. It, too, can then be written like (A.39).

Suppose we now consider two vectors $\underline{U}$ and $\underline{W}$. Something clever happens when I write down their dot product in Cartesian coordinates.

$$\underline{U} \bullet \underline{W} = (U_x\hat{\underline{x}} + U_y\hat{\underline{y}} + U_z\hat{\underline{z}}) \bullet (W_x\hat{\underline{x}} + W_y\hat{\underline{y}} + W_z\hat{\underline{z}})$$
$$= U_xW_x\hat{\underline{x}} \bullet \hat{\underline{x}} + U_xW_y\hat{\underline{x}} \bullet \hat{\underline{y}} + U_xW_z\hat{\underline{x}} \bullet \hat{\underline{z}}$$
$$+ U_yW_x\hat{\underline{y}} \bullet \hat{\underline{x}} + U_yW_y\hat{\underline{y}} \bullet \hat{\underline{y}} + U_yW_z\hat{\underline{y}} \bullet \hat{\underline{z}}$$
$$+ U_zW_x\hat{\underline{z}} \bullet \hat{\underline{x}} + U_zW_y\hat{\underline{z}} \bullet \hat{\underline{y}} + U_zW_z\hat{\underline{z}} \bullet \hat{\underline{z}} \qquad (A.40)$$
$$= U_xW_x\hat{\underline{x}} \bullet \hat{\underline{x}} + U_yW_y\hat{\underline{y}} \bullet \hat{\underline{y}} + U_zW_z\hat{\underline{z}} \bullet \hat{\underline{z}}$$
$$= U_xW_x + U_yW_y + U_zW_z,$$

since all other terms are zero by virtue of (A.37). I have also used (A.38) in the last line. Thus, if I know the x, y, and z-components of any two vectors, I can work out their dot product without any messing around with cosines. Indeed, I do just that when working out the position of the sun from sunrise and sunset data.

Incidentally, Pythagoras' theorem follows trivially from (A.31) and (A.40). Let $\underline{U} = \underline{W}$. Then

$$\underline{W} \bullet \underline{W} = |\underline{W}||\underline{W}|\cos(0)$$
$$= |\underline{W}|^2$$
$$= W_xW_x + W_yW_y + W_zW_z \qquad (A.41)$$
$$= W_x^2 + W_y^2 + W_z^2.$$

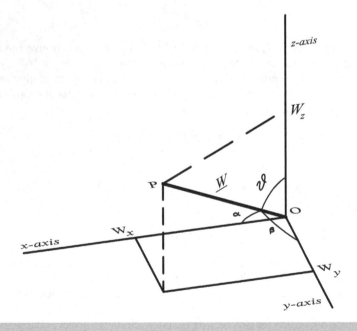

Fig. A.16. Extension of the Cartesian notation for vectors to 3D. The angles θ, α and β are the angles beween the vector $\underline{W}$ and the axes. Since θ is one of the angles used to define spherical polar co-ordinates, i thought it silly to rename it gamma.

Lines two and four are precisely Pythagoras' theorem. This proof is much shorter and neater than the one in my school mathematics textbook, which I think was the one due to Euclid.[42] Indeed, a major benefit of vector notation is that it simplifies proving theorems. Another very useful triangle theorem when working out distances to superior planets is the so-called cosine rule. In Fig. A.16 and (A.29),

$$\underline{E} = \underline{F} - \underline{D}. \tag{A.42}$$

I am going to take the dot product of each side of (A.42) with itself. I then get

$$\begin{aligned} \underline{E} \bullet \underline{E} &= (\underline{F} - \underline{D}) \bullet (\underline{F} - \underline{D}) \\ &= \underline{F} \bullet \underline{F} + \underline{D} \bullet \underline{D} - 2\underline{D} \bullet \underline{F}, \end{aligned} \tag{A.43}$$

i.e.,

$$|\underline{E}|^2 = (\underline{F} - \underline{D}) \bullet (\underline{F} - \underline{D}) = |\underline{F}|^2 + |\underline{D}|^2 - 2|\underline{F}||\underline{D}|\cos(\theta), \tag{A.44}$$

where θ is the angle between $\underline{D}$ and $\underline{F}$. This theorem, proved in four lines flat, is known as the cosine rule. It is used to find the length of the third side of a triangle when you know the lengths of the other two and the angle between them.

I think that is all you need to know about vectors to understand this book.

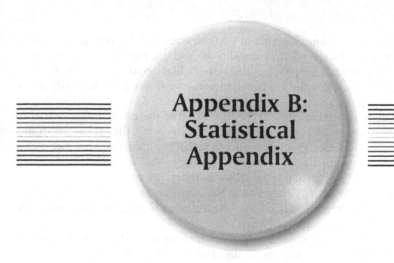

Appendix B:
Statistical
Appendix

This appendix is coming from a different place from the geometrical one. There, I assumed you had merely forgotten nearly everything you once knew about the subject. Here I assume you were much less well trained. Furthermore, I am not very impressed by the way statistics is taught. There is too much emphasis on rote learning techniques and not enough on where the techniques come from. The result is that people make statistical inferences they do not understand. It does not take great prescience to imagine that this could lead to trouble. Indeed, while you might be forgiven for thinking that all branches of mathematics have a bad reputation for being difficult, the only one with a bad reputation for *effectiveness* is statistics.

There is actually a reason for a lot of the rote learning: you need a background in calculus and analysis to understand the derivations of the statistical formulae. Once you have the background, the statistics I want to show you are not especially difficult.

Nevertheless, the reasoning behind the justification for using them is not trivial, and there is no point pretending it is.

If you are not very comfortable with mathematics, you could miss out the section "All Distributions Lead to a Normal Distribution" on a first reading and come back to it later.

The only way I can think of to fit this material into an Appendix is to label some parts of the story as advanced topics; and to warn you that you will need the kind of training someone with a science, engineering, or economics major would have.

The alternative is that I write a book called *A Down to Earth Guide to Statistics for Amateur Astronomers. . .*

Probability

You can make this subject complicated, but I do not want to. I do not think I need to. I think we need two ideas of what probability is. You are meant to imagine some event where we know it can happen, but also view its likelihood as governed in some way by chance. We all know the kinds of phenomena: drawing raffle tickets from hats, etc.

You could have a theoretical view of the likelihood of the event, and state that if you believe that an event can occur in m out of n possible ways, the probability of the event occurring is m/n. For example, you might believe that there are two sides a coin could land on, and only one is heads. Then the probability of getting heads is 1/2. Or the probability of getting raffle ticket number 729 out of a book of 1,000 tickets is 1/1,000. By the way, I have snuck in the assumption that raffle ticket draws are independent. If ticket number 64 has already been drawn, the probability of getting number 729 is no longer 1/1,000.

You can also take an observational approach, and say that in n repetitions of an experiment, you found that your event occurred m times, then you could say that the *empirical probability* of your event is m/n.

You can compare empirical against theoretical probabilities, for example to see if you get swindled the next time you visit the gaming tables at the City of Lost Wages in Nevada. A famous statistician, Karl Pearson, did just that in Monte Carlo, and guess what he found?[43]

Since m cannot be greater than n, or less than zero if it comes to that, we can say that probabilities such as m/n lie anywhere from zero through one. They are sometimes expressed as percentages from 0 to 100%.

Sometimes we can work out theoretical values of m by counting the number of ways n attempts could yield a given result. Let me go back to my raffle tickets to illustrate. At the first draw, there are 1,000 tickets. At the second there remain 999 tickets; at the third there are 998 left, etc. How many ways are there that I could draw, say, three tickets? The answer is that there are $1,000 \times 999 \times 998$ ways. We can use *factorial* notation to write this number down. Remember that n *factorial*, written $n!$, is equal to

$n(n-1)(n-2)(n-3)\ldots 3 \times 2 \times 1$. Therefore

$$100 \times 99 \times 98 = \frac{100 \times 99 \times 98 \times \cdots \times 3 \times 2 \times 1}{97 \times 96 \times 95 \times \cdots \times 3 \times 2 \times 1} = \frac{100!}{97!}. \tag{B.1}$$

It is always true that there are $n!/(n-m)!$ ways of randomly choosing m out of n events. For our raffle tickets, one way we could select the tickets is 64, 729, 1.

But we might also select 1, 64, and then 729. If the first prize is the best prize the order in which we draw these tickets matters. If all the prizes were the same, it would not matter. In that event, which is like a lot of real astronomical situations, all I want to know is about selecting those three tickets in any order. There are three ways you could pick the first ticket (1, 64, and 729). We do not at this point know which ticket was picked first, so there are three different combinations of second ticket (1 & 64, 64 & 729, and 729 & 1). For each of those there are two choices of second ticket. There is only one third remaining ticket for each possible combination, so there are in total

$3 \times 2 \times 1 = 3!$ ways to pick our three tickets. That is always true: there are $m!$ ways to select m tickets. If I am only interested in the number of ways I could get tickets 1, 64, & 729 but do not care about the order, (B.1) overcounts.

Therefore, combining the discussion of the two preceding paragraphs, the number of ways to pick m tickets out of n is

$$\text{Ways to pick } m \text{ tickets out of } n = \frac{n!}{m!(n-m)!}. \tag{B.2}$$

In (B.2), the $m!$ appears in the denominator to divide out all the overcounting. If you do not understand this formula, try a few real examples, with smallish numbers to keep the arithmetic easy, until you do understand it.

A Measurement that Fluctuates

This is really what we need to understand for astronomy. Real measurements are imprecise for many reasons. The causes might be anything from atmospheric turbulence to thermal noise in digital camera chips. Leaving aside the vexed question of whether* "God plays dice with the universe," it is impractical to track down every last cause of variability in measurement. It is, however, practical to *assume* that the variability is random, and use statistical models of this randomness to analyze the variability, and try and obtain a better estimate of the "true" value of a "perfect" measurement. It used to be thought that such a value existed, before the waters were muddied by Heisenberg's discovery of the uncertainty principle. We now believe the concepts of true value and perfect measurement to be approximations.

So, we end up with a simplified conceptual model of our measurement in which we take it that there is a "correct" answer, which we cannot find because of random fluctuations in the values we measure. What we often find is that the more measurements we take, the more constant their average becomes. The fluctuations about the average also have a tendency to settle down to a predictable pattern.

Apart from metaphysics and Heisenberg, there is another whopping great assumption in our conceptual model of a measurement. We assume that there are no *systematic errors.*

What is a systematic error? For example, I discovered a systematic error this winter when I was trying to measure the orbits of Saturn's satellites Titan and Rhea. To take the photos, I was using two Barlow lenses, a $2\times$ for when Titan was far from Saturn and a $4\times$ when it was near. Then there were some strange quirks in the results. In desperation, I decided to do some magnification tests on my Barlow lenses. Guess what? The magnifications were nearer $2.3\times$ and $3.7\times$. It was only a chance reading on the Web that Barlow lenses will work as zoom lenses that led me to make this test. I could indeed make my Barlows work as zoom lenses, by not

---

*Einstein had a tendency to toss off remarks – don't we all? – to which people attached more significance than he probably intended. There was a regrettable tendency to treat every word of his as holy writ. This particular remark has all the hallmarks of bad science, as I am sure he knew. It is simply an assertion, which had no justification in the then available experimental evidence.

pushing the camera all the way into the Barlow lens, and refocusing. Once I corrected my measured distances for this effect, they looked much more believable. This example is typical of systematic errors. They are hard to find, and usually make you look stupid when you do find them, since hindsight is always 20/20.

If measurements vary in an apparently random way, the fluctuation is referred to as a *random error*. The word "error" in this context does not mean "mistake" but it comes from the Latin word "errare" meaning "to wander." Thus random errors are random wanderings from the actual value.

I propose to illustrate how the *normal distribution*, also called the *Gaussian distribution* and the *bell curve*, arises by showing you an example you can generate for yourself with Microsoft Excel or OpenOffice Calc.

I ask you to imagine a measurement of some quantity whose average value over many measurements is 42 units. It does not matter what units we use. This is a theoretical example. We do not know the "correct" value. All we can ever find is the average value. In some statistics texts this average is referred to as the *expectation value*. This example is contrived to make it work in Microsoft Excel. I subject each measurement to 40 random perturbations, each of which might increase or decrease the value I measure by one unit. In Microsoft Excel, I use the random number function RAND( ) and the IF facility. The random number function fills the spreadsheet cell with a random number between 0 and 1. Hence, the typing =IF (RAND( )<0.5,−1,1) will put −1 in the cell if RAND( ) is less than 0.5, and +1 in the cell if RAND( ) is not less than 0.5. Half the time it will insert −1; the other half it will insert +1. In OpenOffice Calc the corresponding function is =IF(RAND( )< 0.5;−1;1) with semicolons instead of commas.

Excel will let me add IF(RAND( )<0.5,−1,1) to 42 forty times before it complains my command line is too long. Hence each

$$\text{Measurement } M = 42 + IF\,(RAND(\,),-1,1) + IF(RAND(\,),-1,1)$$
$$+ \cdots + IF(RAND(\,),-1,1)$$

$$= 42 + \sum_{1}^{40} IF(RAND(\,),-1,1) = 42 + W(k,n), \qquad (B.3)$$

where the Greek sigma symbol $\sum_{1}^{40}$ means sum from 1 to 40, i.e., add the IF(RAND ( )<0.5,−1,1) term 40 times. What happened? Figures B.1–B.3 tell us.

There is no good reason for choosing 287 measurements–it was the first number that came into my head. My wife asked me why I chose it. I think she would be better asking a psychologist. I will be that no one would have questioned me if I had chosen a round number of results like 250 or $256 = 2^8$.

If we look at the first 287 measurements, the distribution of measurements is shown in Fig. B.1. The distribution looks a bit like a bell curve (see also 4.14), but not much.

If we look at the first 2887 measurements (a number also picked for no good reason), the distribution of measurements is shown in Fig. B.2. The distribution looks more like a bell curve, with much less "noise," but it shows some asymmetry about the average value, 42.

If we look at the first 28,887 measurements (a number picked for no better reason than 287 or 2,887), the distribution of measurements is shown in Fig. B.3. The distribution now looks very symmetric about the average value, 42; and

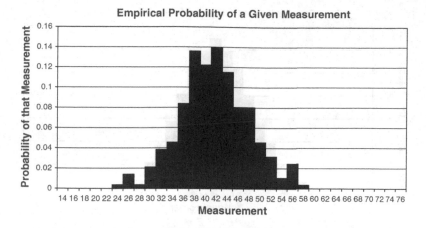

Fig. B.1. Distribution of measurements of our quantity obtained from 287 instances of (B.3).

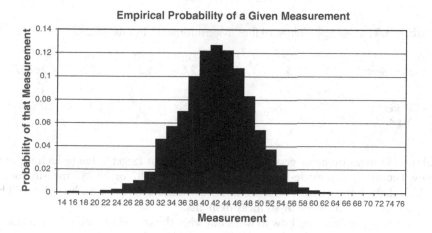

Fig. B.2. Distribution of measurements of our quantity obtained from 2,887 instances of (B.3).

looks like the well-known bell curve. But 28,887 is an awfully large number of measurements.

Why take all these measurements? I am trying to show you the concept of *reproducibility*. This concept deals with the question of whether many repeats of a measurement with random fluctuations tend to give the same result.

Table B.1 shows the average measurements under various circumstances. Taking the average of five measurements does not produce a consistent answer. We say that five measurements are not enough to make our experimental measurement *reproducible*. Taking ten measurements gives an answer within 10% of what we expect. Taking 27 gets us to within 3%. A law of diminishing returns now begins to apply.

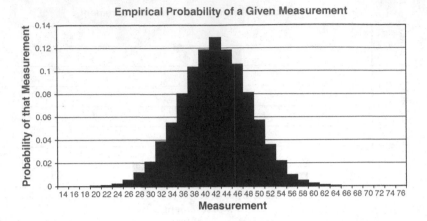

Fig. B.3. Distribution of measurements of our quantity obtained from 28,887 instances of (B.3).

Table B.1. Average Values of the "Measurement" in Our Model

Which Measurements	First 5	Second 5	First 10	First 27	First 287	First 2887	First 28887
% Difference from 42.0	44.8 7%	47.6 13%	46.2 10%	43.3 3%	41.6 −1%	42.0 0%	42.0 0%

Taking 287 measurements gets us to within 1% or out target value (which we only know because it is a made-up example). Going to 2,887 or 28,887 measurements yields a slight improvement for a lot more effort. Most of the reproducibility will be revealed by the first dozen or so measurements.

I now want to show you how to work out what the spread of measurements looks like. Can we work out the characteristics of the distributions of measurements in Fig. B.2 and B.3? The answer is yes. I follow here the treatment of a very similar problem given by one of the few people to win a Nobel Prize for astrophysics, S. Chandrasekhar,[44] although I promise you that this particular calculation is not going to need Nobel-level skills.

The probability of a given fluctuation being positive (or negative) in our example is 1/2. The probability of the next fluctuation being positive is also 1/2. Therefore the probability of both fluctuations being positive is 1/2 of 1/2, i.e., $(1/2)^2$. The probability of a positive step followed by a negative step is also $(1/2)^2$, and so on. After the 40 steps in (B.3), the probability of a given sequence of positive and negative fluctuations is $(1/2)^{40}$. In general, let the result of the fluctuations be a measurement of $42+k$, where k is an integer or whole number. It can be positive, negative or zero. I need to come clean about something I glossed over: in 40 fluctuations from a starting value of 42, the final value has to be even. Every time you add +1 or −1 an even number of times to 42, you get an even number. So k must be even.

Anyway, of the 40 fluctuations, $(1/2)(40+k)$ must have been positive fluctuations and $(1/2)(40-k)$ must have been negative. You can convince yourself of this by trying specific examples. For example, if $k = +2$, there must have been 21 positive fluctuations and 19 negative ones. If $k = +10$, there must have been 25 positive fluctuations and 15 negative ones. If $k = -10$, there must have been 15 positive fluctuations and 25 negative ones, and so on.

We now use (B.2) to work out the number of ways to get to $42+k$ in 40 fluctuations. The probability of this happening is the number of sequences of fluctuations with $(1/2)(40+k)$ positive fluctuations and $(1/2)(40-k)$ negative ones times the probability of each occurring. Let me write this number as $W(k, 40)$. then

$$W(k, 40) = \frac{40!}{[\frac{1}{2}(40 + k)]! \, [\frac{1}{2}(40 - k)]!} \left(\frac{1}{2}\right)^{40}. \tag{B.4}$$

If instead of 40 fluctuations, we have n fluctuations,

$$W(k, n) = \frac{n!}{[\frac{1}{2}(n + k)]! \, [\frac{1}{2}(n - k)]!} \left(\frac{1}{2}\right)^{n}. \tag{B.5}$$

When n is large and $k = n$, we can write an approximate version of this formula. We use an approximation known as Stirling's approximation. I am afraid that the derivation of Stirling's approximation requires rather more mathematical knowledge than the minimum I have assumed, so I will not give it here. At the time of writing, you can find the derivation of this formula on Wikipedia at *http://en.wikipedia.org/wiki/Stirling%27s_formula*, or in the book by Mathews and Walker.[45] Stirling's formula is

$$n! \approx \sqrt{2\pi n} \left(\frac{n}{e}\right)^{n} \tag{B.6}$$

where e is the base of natural logarithms. If you do not like my producing Stirling's formula out of a hat, I do not blame you. I have just done something I have never done before, which is to draw up a table in Microsoft Excel comparing the values of Stirling's approximation with $n!$. Even without the derivation, you can see from Table B.2 that for $n > 9$, Stirling's formula is good to better than 1%.

The main practical advantage of Stirling's formula is that $n!$ overloads electronic calculators because it increases so rapidly as n increases. Stirling's formula is particularly useful in its logarithmic form,

$$\log_e (n!) = \log_e \left(\sqrt{2\pi n}\right) + \log_e ((n/e)^n)$$

$$= \frac{1}{2} \log_e (2\pi) + \frac{1}{2} \log_e (n) + n \log_e(n/e)$$

$$= \frac{1}{2} \log_e (2\pi) + \left(n + \frac{1}{2}\right) \log_e (n) - n \log_e(e)$$

$$= \frac{1}{2} \log_e (2\pi) + \left(n + \frac{1}{2}\right) \log_e (n) - n, \tag{B.7}$$

since the natural logarithm of e is one.

Table B.2. Comparison of $n!$ with Stirling's Formula

N	n!	Stirling's Formula	Ratio
1	1	0.922137009	1.084438
2	2	1.919004351	1.042207
3	6	5.836209591	1.028065
4	24	23.50617513	1.021008
5	120	118.019168	1.016784
6	720	710.0781846	1.013973
7	5,040	4,980.395832	1.011968
8	40,320	39,902.39545	1.010466
9	362,880	359,536.8728	1.009298
10	3,628,800	3,598,695.619	1.008365
20	2.43×10^{18}	2.42279×10^{18}	1.004175
30	2.65×10^{32}	2.64517×10^{32}	1.002782
40	8.16×10^{47}	8.14217×10^{47}	1.002085
50	3.04×10^{64}	3.03634×10^{64}	1.001668
60	8.32×10^{81}	8.30944×10^{81}	1.00139
80	7.2×10^{118}	7.1495×10^{118}	1.001042
100	9.3×10^{157}	9.3248×10^{157}	1.000834
150	5.7×10^{262}	5.7102×10^{262}	1.000556
170	7.3×10^{306}	7.2539×10^{306}	1.00049

Let me apply (B.7) to (B.5). The logarithm of (B.5) is

$$
\begin{aligned}
\log_e \left(W(k,n) \right) &= \log_e (n!) - \log_e \left(\left[\frac{1}{2}(n+k) \right]! \right) - \log_e \left(\left[\frac{1}{2}(n-k) \right]! \right) + n \log_e \left(\frac{1}{2} \right) \\
&\approx \left[\left(n + \frac{1}{2} \right) \log_e(n) - n \right] - \left[\left(\frac{1}{2}(n+k+1) \right) \log_e \left(\frac{1}{2}(n+k) \right) - \frac{1}{2}(n+k) \right] \\
&\quad - \left[\left(\frac{1}{2}(n-k+1) \right) \log_e \left(\frac{1}{2}(n-k) \right) - \frac{1}{2}(n-k) \right] + n \log_e \left(\frac{1}{2} \right) \\
&\quad + \left(\frac{1}{2} - \frac{1}{2} - \frac{1}{2} \right) \log_e (2\pi) \\
&= \left[\left(n + \frac{1}{2} \right) \log_e(n) - n \right] - \left[\left(\frac{1}{2}(n+k+1) \right) \log_e \left(\frac{n}{2} \left(1 + \frac{k}{n} \right) \right) - \frac{1}{2}(n+k) \right] \\
&\quad - \left[\left(\frac{1}{2}(n-k+1) \right) \log_e \left(\frac{n}{2} \left(1 - \frac{k}{n} \right) \right) - \frac{1}{2}(n-k) \right] - n \log_e(2) - \frac{1}{2} \log_e(2\pi) \\
&= \left(n + \frac{1}{2} \right) \log_e(n) - n - \left(\frac{1}{2}(n+k+1) \right) \log_e \left(\frac{n}{2} \left(1 + \frac{k}{n} \right) \right) \\
&\quad + \frac{1}{2}(n+k) - \left(\frac{1}{2}(n-k+1) \right) \log_e \left(\frac{n}{2} \left(1 - \frac{k}{n} \right) \right) + \frac{1}{2}(n-k) \\
&\quad - n \log_e(2) - \frac{1}{2} \log_e(2\pi) \\
&= \left(n + \frac{1}{2} \right) \log_e(n) - \left(\frac{1}{2}(n+k+1) \right) \log_e \left(\frac{n}{2} \left(1 + \frac{k}{n} \right) \right) \\
&\quad - \left(\frac{1}{2}(n-k+1) \right) \log_e \left(\frac{n}{2} \left(1 - \frac{k}{n} \right) \right) - n \log_e(2) - \frac{1}{2} \log_e(2\pi). \quad \text{(B.8)}
\end{aligned}
$$

We can simplify (B.8) further by expanding some of the logarithms as series. This is a slightly more advanced technique, fully explained and derived in many mathematics textbooks, among which my favorite is an old classic by Courant.[46] If you are not familiar with this expansion, you can always test it in a spreadsheet program for various values of k/n.

The series expansion of a logarithm is

$$\log_e\left(1 \pm \frac{k}{n}\right) \approx \pm \frac{k}{n} \pm \frac{k^2}{2n^2} \pm \cdots \tag{B.9}$$

This is valid so long is $k < n$. We substitute (B.9) into (B.8) to yield

$$\log_e(W(k,n)) \approx \left(n + \frac{1}{2}\right)\log_e(n) - n\log_e(2)$$

$$- \left[\left(\frac{1}{2}(n+k+1)\right)\log_e\left(\frac{n}{2}\left(1 + \frac{k}{n}\right)\right)\right]$$

$$- \left(\frac{1}{2}(n-k+1)\right)\log_e\left(\frac{n}{2}\left(1 - \frac{k}{n}\right)\right) - \frac{1}{2}\log_e(2\pi)$$

$$= \left(n + \frac{1}{2}\right)\log_e(n) - \frac{1}{2}\log_e(2\pi) - n\log_e(2)$$

$$- \left(\frac{1}{2}(n+k+1)\right)\left[\log_e(n) - \log_e(2) + \log_e\left(1 + \frac{k}{n}\right)\right]$$

$$- \left(\frac{1}{2}(n-k+1)\right)\left[\log_e(n) - \log_e(2) + \log_e\left(1 - \frac{k}{n}\right)\right]$$

$$\approx \left(n + \frac{1}{2}\right)\log_e(n) - \frac{1}{2}\log_e(2\pi) - n\log_e(2)$$

$$- \left\{\left(\frac{1}{2}(n+k+1)\right)\left[\log_e(n) - \log_e(2) + \frac{k}{n} - \frac{k^2}{2n^2} + \cdots\right]\right\}$$

$$- \left(\frac{1}{2}(n-k+1)\right)\left[\log_e(n) - \log_e(2) - \frac{k}{n} - \frac{k^2}{2n^2} + \cdots\right]$$

$$= \left(n - \frac{n}{2} - \frac{n}{2} + \frac{1}{2} - \frac{1}{2} - \frac{1}{2}\right)\log_e(n) - \frac{1}{2}\log_e(2\pi)$$

$$+ \left(-n + \frac{n}{2} + \frac{n}{2} + \frac{1}{2} + \frac{1}{2}\right)\log_e(2) - \left(\frac{1}{2}(n+k+1)\right)\left[\frac{k}{n} + \frac{k^2}{2n^2}\right]$$

$$- \left(\frac{1}{2}(n-k+1)\right)\left[-\frac{k}{n} - \frac{k^2}{2n^2}\right]$$

$$\approx \left(-\frac{1}{2}\right)\log_e(n) - \frac{1}{2}\log_e(2\pi) + \log_e(2) - \left(\frac{1}{2} + \frac{1}{2}\right)\frac{k^2}{2n} + \cdots$$

$$= \frac{-1}{2}\log_e(n) - \frac{1}{2}\log_e(2\pi) + \log_e(2) - \frac{k^2}{2n}$$

$$= \frac{1}{2}\log_e\left(\frac{2}{\pi n}\right) - \frac{k^2}{2n}. \tag{B.10}$$

We are almost there. Taking the antilogarithm of (B.10) yields

$$W(k, n) = \sqrt{\frac{2}{\pi n}} \exp\left(\frac{-k^2}{2n}\right). \tag{B.11}$$

If k were a continuous variable, not an integer, (B.11) would be the equation of a normal or Gaussian distribution. As it is, k can only take certain values, but the distribution of $W(k, n)$ is a normal distribution.

But, you rightly ask, was not this a rather contrived example? A Mickey Mouse one even? Yes it was. The remarkable thing is that the same thing happens for virtually any distribution of random errors.

All Distributions Tend Toward a Normal Distribution

I am now going to take you through the proof of this remarkable claim: *all distributions of random fluctuations tend toward normal distributions if there are enough fluctuations.* I will have to meander somewhat to set up some results to use in my proof. Your patience is requested.

Here is something else worth asking about the distribution $W(k, n)$ of measurements. How scattered are they? The way statisticians usually get a handle on this is to think not only in terms of the average or mean of a measurement, but also in terms of its *variance*.

To define this, we need to do some work on the mean of $W(k, n)$. I do not know why but you have more street cred if you say "mean" rather than "average." Let me call it μ (Greek lower case "mu"). It is obvious from Fig. B.1 through Fig. B.3 that, in terms of (B.3), the mean value of M is 42, and the mean of $W(k, n)$ is therefore zero. The same conclusion could quickly be drawn from (B.11), where it can be seen that $W(k, n) = W(-k, n)$ because $W(k, n)$ only depends on k^2. It would therefore be absurd for the mean value of $W(k, n)$, μ, to be anything other than zero.

A more mathematical definition of the mean[47] is this:

$$\mu = \sum_k kW(k, n), \tag{B.12}$$

where the sum is taken over all possible values of k. Since $W(k, n) = W(-k, n)$, the value of μ in (B.12) has to be zero because every term $k\,W(k, n)$ is canceled out by a term $-kW(-k, n) = -kW(k, n)$. In general for other functions $W(k, n)$, μ need not be zero. If k were continuous, i.e., could have noninteger values, the mean would be defined[48] as

$$\mu = \int kW(k, n)dk, \tag{B.13}$$

where the integral is really a definite integral taken over all possible values of k.

Back to variance. Note that $(k-\mu)$ could be positive or negative, but $(k-\mu)^2$ is always positive. Out of left field, let me ask: what is the mean of $(k-\mu)^2$? The answer, using the language of (B.12), is

$$\sigma^2 = \sum_k (k - \mu)^2 W(k, n). \tag{B.14}$$

This is in fact the definition of variance that statisticians use. If k were continuous, the definition of variance would be

$$\sigma^2 = \int (k - \mu)^2 W(k, n) dk, \tag{B.15}$$

where again the integral is really a definite integral taken overall possible values of k.

The variance is given the name σ^2, not σ (Greek lower case "sigma") because its square root σ also has a name: it is called the *standard deviation* of the distribution. That is all I am going to tell you about variance and standard deviation for now. This is not meant to be a treatise on statistics, merely an Appendix to get you through the main chapters of the book.

I am sorry to skip around and introduce apparently mad ideas. Please trust me. I am getting there. I do need to tell you about something called a *moment generating function*. This is a rather scary name for a function which really is not very complicated. It is a bit like a mean only different. Its definition is

$$G(t) = \sum_k \exp(t\,k) W(k, n) - \text{discrete } k,$$

$$G(t) = \int \exp(t\,k) W(k, n) dk - \text{continuous } k. \tag{B.16}$$

In fact by analogy with (B.12) and (B.13), $G(k)$ is the mean of $\exp(t\,k)$.

We exploit the fact that

$$\exp(t\,k) = 1 + (t\,k) + \frac{(t\,k)^2}{2!} + \frac{(t\,k)^3}{3!} + \frac{(t\,k)^4}{4!} + \cdots \tag{B.17}$$

to write

$$G(k) = \sum_k W(k, n) + \sum_k (t\,k) W(k, n) + \sum_k \frac{(t\,k)^2}{2!} W(k, n) + \sum_k \frac{(t\,k)^3}{3!} W(k, n)$$

$$+ \cdots - \text{discrete } k,$$

$$G(k) = \int W(k, n) dk + \int (t\,k) W(k, n) dk$$

$$+ \int \frac{(t\,k)^2}{2!} W(k, n) dk + \int \frac{(t\,k)^3}{3!} W(k, n) dk + L - \text{continuous } k. \tag{B.18}$$

You may care to note that the second term in each of the two versions of (B.18) looks like our mean ((B.12) and (B.13)), and the third term bears a passing resemblance to our variance ((B.14) and (B.15)). Note also that for every different form of $W(k, n)$, there can only be one moment generating function. The converse is true. The moment generating function for $W(k, n)$ cannot be the moment generating function for any other probability distribution.

Oh, and I have to confess to having contrived my example so that the mean of $W(k, n)$ came out to zero. That is a common trick in statistics: to subtract the mean value from a random variable to keep mathematical manipulation simple. You can make it even simpler if you make the following transformation of a random variable k.

$$Z = \frac{k - \mu}{\sigma}. \tag{B.19}$$

Z then has a mean of zero and a variance of one. In some texts you may see it referred to as a *standardized random variable*.

Now for another meander. Suppose I have a load of sets of random variables X_i where $i = 1,2,3,\ldots,n$. Suppose further that each has the same mean μ and variance σ^2. Now let me, just for the heck of it, add 'em all up.

$$S_n = X_1 + X_2 + X_3 + \cdots + X_n = \sum_{i=1}^{n} X_i. \tag{B.20}$$

The mean of S_n is therefore

$$\text{Mean of } S_n = \sum_{i=1}^{n} \mu = n\mu. \tag{B.21}$$

The variance of S_n is

$$\text{Var}(S_n) = \sum_{i=1}^{n} \text{Var}(X_i) = \sum_{i=1}^{n} \sigma^2 = n\sigma^2. \tag{B.22}$$

The standardized random variable for S_n, per (B.19), is therefore

$$S_n^* = \frac{S_n - n\mu}{\sqrt{n}\sigma}. \tag{B.23}$$

So [expletive, deleted] what, you ask? I am getting there. I am getting there.

The moment generating function for S_n^* is

$$\begin{aligned}
G(S_n^*) &= \text{Mean}\left(\exp(S_n^* t)\right) \\
&= \text{Mean}\left(\exp\left(\frac{S_n - n\mu}{\sigma\sqrt{n}} t\right)\right) \\
&= \text{Mean}\left(\exp\left(\frac{X_1 - \mu}{\sigma\sqrt{n}} t\right)\exp\left(\frac{X_2 - \mu}{\sigma\sqrt{n}} t\right)\cdots\exp\left(\frac{X_n - \mu}{\sigma\sqrt{n}} t\right)\right) \\
&= \left(\text{Mean}\left(\exp\left(\frac{X_1 - \mu}{\sigma\sqrt{n}} t\right)\right)\right)^n \\
&= G(X_n)^n,
\end{aligned} \tag{B.24}$$

since each of the random variables X_i has the same μ and σ. By a series expansion of the exponential function

$$G_{X_n}(t) = \text{Mean}\left(\exp\left(\frac{X_1 - \mu}{\sigma\sqrt{n}}t\right)\right)$$

$$= \text{Mean}\left(1 + \frac{X_1 - \mu}{\sigma\sqrt{nt}} + \frac{(X_1 - \mu)^2}{2\sigma^2 n}t^2 + \cdots\right)$$

$$= \text{Mean}(1) + \frac{\text{Mean}(X_1 - \mu)}{\sigma\sqrt{n}}t + \frac{\text{Mean}(X_1 - \mu)^2}{2\sigma^2 n}t^2 + \cdots \qquad \text{(B.25)}$$

$$= 1 + \frac{0}{\sigma\sqrt{n}}t + \frac{\sigma^2}{2\sigma^2 n}t^2 + \cdots$$

$$= 1 + \frac{\sigma^2}{2\sigma 2n}t^2 + \cdots$$

Therefore

$$G_{S_n^*}(t) = G_{X_n}(t)^n = \left(1 + \frac{\sigma^2}{2\sigma^2 n}t^2 + \cdots\right)^n = \left(1 + \frac{t^2}{2n} + \cdots\right)^n. \qquad \text{(B.26)}$$

Now, the limit of (B.26) as n tends to infinity is

$$\lim_{n\to\infty} G_{S_n^*}(t) = \lim_{n\to\infty}\left(1 + \frac{t^2}{2n} + \cdots\right)^n = \exp\left(\frac{t^2}{2n}\right)^n = \exp\left(\frac{t^2}{2}\right). \qquad \text{(B.27)}$$

A little earlier I claimed that all probability distributions tend to the normal distribution. If I am right, (B.27) should be equal, or at the very least proportional, to the moment generating function for a normal distribution. Let us check. Well, let me calculate the moment generating function for a normal distribution

$$f(x) = \frac{1}{\sigma\sqrt{2\pi}}\exp\left(\frac{-x^2}{2\sigma^2}\right). \qquad \text{(B.28)}$$

Then

$$G(t) = \frac{1}{\sigma\sqrt{2\pi}}\int_{-\infty}^{\infty}\exp(tx)\exp\left(\frac{-x^2}{2\sigma^2}\right)dx$$

$$= \frac{1}{\sigma\sqrt{2\pi}}\int_{-\infty}^{\infty}\exp\left(\frac{-x^2}{2\sigma^2} + tx\right)dx$$

$$= \frac{1}{\sigma\sqrt{2\pi}}\int_{-\infty}^{\infty}\exp\left(-\frac{x^2 - 2t\sigma^2 x}{2\sigma^2}\right)dx$$

$$= \frac{1}{\sigma\sqrt{2\pi}}\int_{-\infty}^{\infty}\exp\left(-\frac{x^2 - 2t\sigma^2 x + (\sigma t)^2 - (\sigma t)^2}{2\sigma^2}\right)dx$$

$$= \frac{1}{\sigma\sqrt{2\pi}}\int_{-\infty}^{\infty}\exp\left(-\frac{(x - t\sigma)^2 - (\sigma t)^2}{2\sigma^2}\right)dx \qquad \text{(B.29)}$$

$$= \frac{\exp(t^2/2)}{\sigma\sqrt{2\pi}}\int_{-\infty}^{\infty}\exp\left(-\frac{(x - t\sigma)^2}{2\sigma^2}\right)dx$$

$$= \frac{\exp(t^2/2)}{\sigma\sqrt{2\pi}}\int_{-\infty}^{\infty}\exp\left(-\frac{x^2}{2\sigma^2}\right)dx$$

$$= \text{Constant} \times \exp(t^2/2),$$

since $\int_{-\infty}^{\infty} \exp(-x^2/(2\sigma^2))dx$ is a constant. In the last line but two of (B.29), I made use of the fact that since x runs from minus infinity to infinity, it makes no difference to the integral if $t\sigma$ is subtracted from x. I could have evaluated the integral in the last line of (B.29), but it is rather paper- and brain-consuming to evaluate this integral; and there is no point. I have proved that the moment generating function for a Gaussian distribution is proportional the one in (B.27). Therefore all distributions tend to one that is proportional to a normal distribution. Any distribution that differs form a normal distribution only by a constant of proportionality must be a normal distribution. So I have proved my point.

What Happens When I Can Only Make One Measurement per Observation of a Moving Body?

In this case, we might end up with a series of observations like, say, those in Fig. B.4. There is a pattern, in this case perhaps a straight line relationship, but each of the measurements is also subject to random fluctuations. By virtue of the above discussion, we can safely assume that the fluctuations at each point would be normally distributed if we had been able to make many observations at that point.

We do not need to. We can exploit the knowledge that such measurements at each point *would* be normally distributed to obtain a *best fit* to the data in Fig. B.4 or any similar data set. Often, our horizontal axis is time. Nowadays we can measure time very accurately and precisely, so we can take it that all the fluctuations occur in the other quantity.

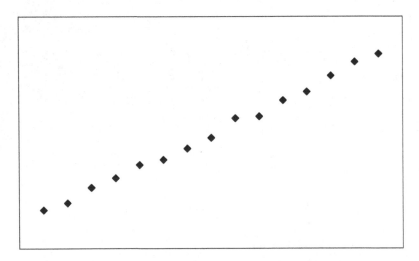

Fig. B.4. A set of measurements of a quantity which shows a behavior pattern, but also some random noise.

We assume, then, that whatever quantity this is, we have measured one of a sample of data points normally distributed about a mean value with probability $\exp(-\Delta y_i^2/2\sigma^2)$, where Δy_i is the variation of the y value of the ith measurement. We further assume that each measurement has the same variance σ^2. Since the same measurement technique is usually used for all the measurements, this is a reasonable assumption; and that each measurement varies independently of the others. Then the probability of obtaining the whole set of measurements

$$P_{set} = \exp\left(-\frac{\Delta y_1^2}{2\sigma^2}\right)\exp\left(-\frac{\Delta y_2^2}{2\sigma^2}\right)\cdots\exp\left(-\frac{\Delta y_n^2}{2\sigma^2}\right)$$

$$= \exp\left(-\frac{\Delta y_1^2 + \Delta y_2^2 + \cdots + \Delta y_n^2}{2\sigma^2}\right) \tag{B.30}$$

the best fit is that which would maximize P_{set}. The lower the value of $\Delta y_1^2 + \Delta y_2^2 + \cdots + \Delta y_n^2$, the higher is P_{set}. Therefore *the best fit occurs when we minimize the sum of the squares of the fluctuations in y.*

This is a really useful result. It works in all sorts of situations, and saves us an awful lot of work.

In a situation like Fig. B.4 where the best fit is a straight line, or indeed a polynomial, it is possible to solve algebraically for the least squares fit. This is done in many textbooks for the straight line case, but the Schaum book on Probability and Statistics also derives the least squares parabola case.[49]

Other cases cannot be so solved, and we have to use our wits.

The method I used was to adapt a so-called Monte Carlo method. The idea was to use a spreadsheet program – Microsoft Excel and OpenOffice Calc should both work – to run a large number of trial *models* of my data, and to choose the one which has the lowest sum of the squares of the fluctuations in *y*. For example, when I was looking at Saturn's moons (results not included in this text) my model of the distance of the moon from the parent planet in the direction of the planet's equator was

$$y = A\cos(\omega t + \varphi), \tag{B.31}$$

where A is an unknown constant, ω is an unknown angular frequency of the orbit, and φ is an unknown phase. In other words I have three unknowns. I tried instances of

$$y = (A_0 + \delta A)\cos((\omega_0 + \delta w)t + (\varphi_0 + \delta\varphi)), \tag{B.32}$$

where δA, $\delta\omega$, and $\delta\varphi$ are random numbers added to A_0, ω_0, and φ_0, respectively. A_0, ω_0, and φ_0 are my initial guesses of the values of A, ω, and φ.

I am going to illustrate this principle via a table. In practice the number of data points is much greater, but I am trying to show you how to do the analysis, not what answers I got.

In Table B.3, I quite arbitrarily set

$$\delta\omega = 0.2\omega_0; \quad \delta A = 0.1A_0; \text{ and } \delta\varphi = 0.2\varphi_0. \tag{B.33}$$

In order to calculate the numbers in the cells in the three leftmost columns, I then used the formulae

$$\frac{2\pi}{\omega_0 + \delta\omega} = 0.95 \times \frac{2\pi}{\omega_0} + 0.10 \times RAND();$$

$$A_0 + \delta A = 0.95 \times A_0 + 0.10 \times RAND(); \text{ and} \tag{B.34}$$

$$\varphi_0 + \delta\varphi = 0.90 \times \varphi_0 + 0.20 \times RAND(),$$

where I have written the formulae in the syntax of Microsoft Excel and OpenOffice Calc. The idea is that, since δA is $0.1A_0$, in (B.33), A can fluctuate from $0.95A_0$ to $1.05A_0$; since $\delta\varphi$ is $0.2\varphi_0$, in (B.34), φ can fluctuate from $0.9\varphi_0$ to $1.1\varphi_0$, and so on.

I was really after the orbital period $2\pi/\omega$ rather than ω. In the fifth through seventh columns I put the calculated values from (B.32). In the eighth through tenth columns, I have worked out the differences Δy_1^2, Δy_2^2, and Δy_3^2 between observed and modeled values; and in the eleventh column I have added them to give $\Delta y_1^2 + \Delta y_2^2 + \Delta y_3^2$. This sum of squares is the very quantity I am trying to minimize. In the final column, I have selected the minimum value of $\Delta y_1^2 + \Delta y_2^2 + \Delta y_3^2$, which is 0.004, which inspection showed came from my seventh trial model (there are ten trial models in Table B.3). Therefore this seventh trial model is my best fit, with

$$A = 3.517; \quad 2\pi/\omega = 16.457; \text{ and } \varphi = 0.955.$$

The orbital period of Titan is actually about 15.945 days, so my value of 16.457 days is not outrageous considering how few data points I have used. When I re-ran the Monte Carlo simulation ten times with the same number of points, I found orbital periods between about 15.3 and 17.3 days. So the answer in (B.35) is a luckier than I really deserve. There are not enough data. When I did it for real, I had 58 measurements taken at different times, and instead of ten trials I had 32,772 trials. As a matter of practical convenience, I had 8,193 trials on the Excel worksheet, and I copied and used paste-special-values to put the data onto four separate worksheets to store it while I recalculated. This is because the RAND( ) function in Excel has the irritating habit of recalculating every time you do something. When you only have one or two of them in a worksheet, it does not matter, but it takes a minute or so to munch through $8,193 \times 58 \times 3 = 1,425,582$ instances of RAND( ), even on my rather fast desktop computer.

The outcome of my Monte Carlo simulation was that the orbital period of Titan was best fit by a value of 15.911 days, which is 0.22% different from the published value. That was quite impressive. However, the method proved to be poor at fitting amplitudes A for Titan and Rhea. By "poor" I mean that the results were no better than $\pm 2\%$. The value of the least squares difference did not change much if A changed. Indeed the only way I could get good fixes on A was to take photos very close to the moment of maximum distance from Saturn, and make measurements from them. For Jupiter's moons, I noted the same phenomenon, but it was not such a difficulty. I still got good fits. What I often did was to run a second Monte Carlo simulation with less fluctuation, using the fits from the first simulation as my starting data. That often got me better fits.

Table B.3. Table from Monte Carlo Analysis of Titan's Orbit

$2\pi/\omega_0$	A_0	φ_0	A_0								
15.945	3.428	0.95	Time since start of experiment (days)	0.242	1.183	1.204	0.242	1.183	1.204		
			Titan equatorial distance (min)	1.761	0.618	0.589	Difference of squares	Difference of squares	Difference of squares	Sum of differences of squares	Least squares fit
$2\pi/(\omega_0 + \delta\omega)$	$A_0 + \delta A$	$\varphi_0 + \delta\varphi$	Trial fit	1.719	0.528	0.500					0.004
17.477	3.381	0.896		1.874	0.834	0.809	0.013	0.046	0.049	0.108	
15.113	3.489	0.895		1.898	0.639	0.609	0.019	0.000	0.000	0.020	
14.727	3.391	1.029		1.440	0.126	0.096	0.103	0.242	0.243	0.588	
14.460	3.483	0.931		1.774	0.436	0.405	0.000	0.033	0.034	0.067	
15.515	3.271	0.915		1.730	0.573	0.546	0.001	0.002	0.002	0.005	
16.373	3.359	1.027		1.462	0.299	0.273	0.089	0.102	0.100	0.291	
16.457	3.517	0.955		1.756	0.573	0.545	0.000	0.002	0.002	0.004	
14.460	3.269	0.871		1.831	0.604	0.575	0.005	0.000	0.000	0.005	
14.599	3.397	0.950		1.678	0.379	0.349	0.007	0.057	0.058	0.122	
17.488	3.314	1.012		1.505	0.441	0.416	0.065	0.031	0.030	0.126	

This data set is a Monte Carlo simulation (of fluctuating data)

Therefore, this method is not perfect, and needs to be taken with a pinch of salt. Do not be gulled into believing your results until you have checked them. In real research situations, of course, you may not have any published data to compare with. You then have to make a decision whether you believe your results. Ask yourself these questions. Does a slight perturbation of the best fit still produce a good fit? If not, you may have an error in your mathematics. Do the second and third best fits give very similar answers to the best one? If you plot your fitted data, do they look like a good fit to the measured data? If not, poke around and try to find out why not. And, above all, leave your results for a day or two and then recheck them. It is amazing how many errors you then find.

Nevertheless it enabled me to work out satellite orbits and the distances to Saturn and Mars, so it is a practical technique if used with care.

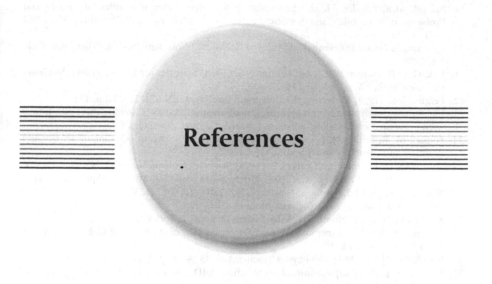

References

1. Tatum, J. B. (2007) Celestial Mechanics, An online book found at http://www.astro.uvic.ca/~tatum
2. Bakich, M. E. (2000) The Cambridge Planetary Handbook, Cambridge University Press, Cambridge, ISBN 9780521632805.
3. Huyghens Christiaan, (1629–95) quoted in Harrison, E. (2000) Cosmology: the Science of the Universe, Second Ed., Cambridge University Press, Cambridge, ISBN 052166148X, p. 54.
4. Newton, I. (1687) Mathematical Principles of Natural Philosophy (Philosophiae Naturalis Principia Mathematica), Translated by Motte A. (1846), First American Edition, New York, Available online at http://rack1.ul.cs.cmu.edu/is/newton
5. Murray, C. D. and Dermott, S. F. (2000) Solar System Dynamics, Cambridge University Press, Cambridge, ISBN: 0521575974.
6. Kepler, J. (1619) The Harmony of the World, Translated by J. Field (1997), The American Philosophical Society, Philadelphia, PA, ISBN 0-87169-209-0.
7. Ptolemy, C. (c. 100) The Almagest, Translated by G. J. Toomer (1998) as "Ptolemy's Almagest", Princeton University Press, Princeton, NJ, ISBN 0-691-00260-6.
8. http://sci.esa.int/science-e/www/object/index.cfm?fobjectid = 38833&fbodylongid = 1860
9. http://77illinois.homestead.com/files/astro/planets3.html
10. http://rathnasree.htmlplanet.com/befriending_venus.htm
11. http://sunearth.gsfc.nasa.gov/eclipse/transit/venus0412.html
12. Bakich, M. E. (2000) The Cambridge Planetary Handbook, work cited.
13. Kaye, G. W. C., Laby, T. H., et al. (1995) Tables of Physical and Chemical Constants, 16th Ed., Longman, Harlow, Essex.
14. French, A. P. (1971) Newtonian Mechanics. Nelson, London.
15. Thomas, G. B. Jr. (1960) Calculus and Analytic Geometry, Third Ed., Addison Wesley, Reading, MA, pp. 437–504.
16. Murray, C. and Dermott, S. F. (1999) Solar System Dynamics, Cambridge University Press, Cambridge, p. 531.
17. French, A. P. (1971) Newtonian Mechanics, work cited, p. 696.
18. United Nations Atlas of the Oceans (searched 2008) http://www.oceansatlas.org/unatlas/about/physicalandchemicalproperties/background/seemore1.html

19. Spiegel, M.R. Schiller, J.J. and Srinivasan, R.A. (2000) Schaum' Outline of Theory and Problems of Probability and Statistics, Second Ed., McGraw-Hill, New York, NY, ISBN 0071350047.

20. Courant, R. (1934) Differential and Integral Calculus, 2 Vols, Republished, Wiley, New York, NY, 1988.

21. Box, G. E. P., Hunter, W. G., and Hunter, J. S. (1978) Statistics for Experimenters, McGraw-Hill, New York, NY, p. 461, Eq. (14–27).

22. Ferguson, K. (1999) Measuring the Universe, Headline, ISBN 0747221324, p. 175.

23. http://www.cloudynights.com/ubbthreads/attachments/1385105-ICX098BQshort.pdf; http://www.unibrain.com/download/pdfs/Fire-i_Board_Cams/ICX098BQ.pdf

24. Spiegel, M. R, Schiller, J. J., and Srinivasan, R. A. (2000) Schaum's Outline of Theory and Problems of Probability and Statistics, work cited.

25. Freedman, D., Pisani, R., and Purves, R. (1978) Statistics, Norton, New York, NY, pp. 192–197.

26. Robert, C. P. and Casella, G. C. (2004) Monte Carlo Statistical Methods (Springer Texts in Statistics), Second Ed., Springer, New York, NY, ISBN-10: 0387212396.

27. http://en.wikipedia.org/wiki/Galilean_moons

28. Murray, C. D. and Dermott, S. F, Solar System Dynamics, work cited.

29. Cohen, I. B. (1942) Roemer and the First Determination of the Speed of Light, The Burndy Library Inc., New York, NY.

30. Stauffer, R. H. Jr. (1997) The Physics Teacher, Vol. 35, April, p. 231.

31. http://store.pasco.com/pascostore/showdetl.cfm?&DID = 9&Product_ID = 1655&groupID = 306&Detail = 1

32. Sheehan, W. (1996) The Planet Mars: A History of Observation and Discovery, University of Arizona Press, Tucson, AZ, ISBN 0816516413.

33. Bakich, M. E. (2000) The Cambridge Planetary Handbook, work cited, p. 49.

34. Covington, M. A. (1999) Astrophotography for the Amateur, Second Ed., Cambridge University Press, Cambridge, ISBN 0521627400.

35. Danby, J. M. A. (1992) Fundamentals of Celestial Mechanics, Second Edition, Willman-Bell, Richmond, VA, ISBN 978-0943396200.

36. Bakich, M. E. (2000) The Cambridge Planetary Handbook, work cited.

37. Walkenbach, J. and Underdahl, B. (2001) Excel 2002 Bible, Wiley, New York, NY, ISBN 0764535838.

38. Tatum, J. B. (2007) Celestial Mechanics, work cited, Chapter 8.

39. Bakich, M. E. (2000) The Cambridge Planetary Handbook, work cited, p. 125.

40. Kaye, G. W. C. and Laby, T. H. (1973) Tables of Physical and Chemical Constants, 14th Edition, Longman, pp. 11 and 129.

41. Bakich, M. E. (2000) The Cambridge Planetary Handbook, work cited.

42. Euclid (c. 300 BCE)The Elements of Euclid,Edited by Todhunter, I., Everyman' Library Edition (1933), Dutton & Co., New York, NY/Dent & Sons, London.

43. Pearson, K., 1900, cited in Wall, J. V. and Jenkins, C. R., 2003, Practical Statistics for Astronomers, Cambridge University Press, Cambridge, ISBN 0521456169, p. 87.

44. Chandrasekhar, S. (1943) Stochastic problems in Physics and Astronomy, *Reviews of Modern Physics*, **15**(1), 1–89.

45. Mathews, J. and Walker, R. L. (1970) Mathematical Methods of Physics, Second Ed., Wiley, New York, NY, ISBN 0805370021, pp. 89–90.

46. Courant, R. (1937) Differential and Integral Calculus, Vol. I, Second Ed., Translated by McShane J. E., Blackie, London, pp. 316–318. This book has been reissued as part of the Wiley Classics Library, but used copies are still available from Amazon.

47. Spiegel, M. R., Schiller, J. J., and Srinivasan, R. A. (2000) Schaum's Outline of Theory and Problems of Probability and Statistics, work cited.

48. Spiegel, M. R, Schiller, J. J., and Srinivasan, R. A. (2000) Schaum's Outline of Theory and Problems of Probability and Statistics, work cited.

49. Spiegel, M. R, Schiller, J. J., and Srinivasan, R. A. (2000) Schaum's Outline of Theory and Problems of Probability and Statistics, work cited.

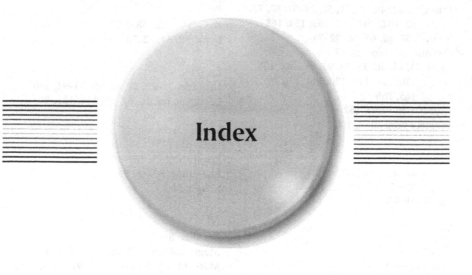

Index